New Technologies and National Security Challenges

Neil Kent, University of Cambridge
Irina du Quenoy, Georgetown University

New Technologies and National Security Challenges

Neil Kent, University of Cambridge
Irina du Quenoy, Georgetown University

Academica Press
Washington - London

Library of Congress Cataloging-in-Publication Data

Names: Kent, Neil, editor | du Quenoy, Irina, editor.
Title: New technologies and national security challenges / Neil Kent, Irina du Quenoy.
Description: Washington : Academica Press, [2020] |

Identifiers: LCCN 2020930355 | ISBN 9781680539981

(hardcover : alk. paper)
LC record available at https://lccn.loc.gov/2020930355

Contents

The Turkish Cyber Security Strategy: Structure, Legislation, and Challenges

Emin Daskin (Ghent University)

Abstract

Given that Turkey enjoys an important geostrategic position, plays multiple critical roles in its region, and is a key NATO member, the country's cyber security has far-reaching implications. As such, an examination of Turkey's cyber security strategy, its defensive and offensive cyber capabilities and the main threats facing the country on this front is warranted. This article gives a brief history of the drafting of Turkey's first cybersecurity strategy, followed by an analysis of the setting-up of the Turkish cybersecurity community and a discussion of its aims and goals. The conclusion highlights the phenomenon of hacktivism as a central challenge to Turkey's cyber security. The article suggests that the country's early enthusiasm for dealing with cyber threats has shrunk and questions whether the adopted strategy has been successfully implemented, in light of recent cyber incidents. The paper also touches upon the effects on Turkish cyber security of political turmoil, infiltration of the state apparatus and agencies by the Gülen sect, and the population's social vulnerability and lack of digital safety awareness.

Introduction

Over the past two decades, Turkey has increased its international ambitions and directed its policies toward being present on all platforms and industries. Bolstered by these efforts, Turkey enjoys an important geostrategic position, plays multiple critical roles in its region, and is a key NATO member. As such, the security of the country's cyberspace has far reaching implications, making it imperative to examine its cyber security strategy, capacity, and the threats faced.

Just as it has with other important initiatives, such as improvements to transportation and health, Turkey attempted to achieve significant progress in the areas of digitalization, cyberspace, and cyber security in a very short time. An ambitious strategy was developed that created new structures and included all possible stakeholders. This was followed initially by an enthusiastic implementation process. Within a few years, however, this ambition and enthusiasm were scaled back to more realistic levels. Large amounts of funding continued to be channeled to R&D projects and the development of academic expertise and training programs, with the main concerns being related to economic and security-related questions.

This article starts by briefly summarizing the efforts that shaped the road to the current strategy. It then examines the legislation and strategy related to cyber incidents and crimes, and the creation of agencies for protective measures and intelligence collection. Next, the structure of Turkey's cyber security landscape is described. The last section of the article illustrates the cyber threats Turkey deals with via a discussion of recent cases of cyberattacks and incidents related to the phenomenon of hacktivism.

Before continuing, it should be noted that the Turkish cyber security strategy, legislation, and the structures involved are very new and thus not fully developed yet. Academic research on this topic is limited, mostly focusing on certain aspects of cybercrime; meanwhile, the few existing publications by government agencies are very repetitive and characterized by a lack of transparency. Finally, the author wishes to make clear that he does not take a stance on any of the discussed cyber security incidents, involved agencies, references, and cited sources; that all data mentioned in this article have been collected exclusively through various open source documents; and that he has no conflict of interest in writing this piece.

The First Efforts to Develop a Strategy

The first time the Turkish government officially considered cyber security on a high level was at a meeting of the National Security Council in 2010, during which the cyber threats toward Turkey and attempts to counter them were discussed (according to the meeting's final report; see

T. C. Milli Güvenlik Kurulu Genel Sekreterligi 2010). Subsequent to these discussions, Turkey's first and so far only cyber security strategy was created on June 19, 2012, during a workshop organized by the Ministry of Transportation, Maritime, and Communications, with the cooperation of the Association for Information Security and the Union of Turkish Bar Associations, the sponsorship of Huawei, and the participation of several government agencies, universities, and private companies (veTeknoloji 2012). In an initial draft, the workshop participants set the strategy's goals, a limited roadmap, and the first steps to be taken to create a cyber security structure. The draft informed the fundamentals of Turkey's cyber security policy and was copied almost word for word in the strategy's publicly accessible official version. As such, it needs to be discussed here in some detail.

The June 2012 draft articulates three main goals, namely: 1) increasing security of Turkish cyber space; 2) assuring durability and continuity of critical infrastructure protection against cyberattacks; and 3) assuring the protection of individual and institutional data. In addition, a set of values was defined that was deemed critical to ensuring the utility and benefits of citizens' use of cyberspace: 1) the protection of basic rights and liberties; 2) compliance with the requirements of democratic society and order; 3) compliance with the rule of law; 4) inclusion of all stakeholders in the process of decision making; 5) finding a balance between privacy, security, and usability; 6) compliance with international regulations; 7) developing a holistic approach toward cyber security, including legal, technical, administrative, economic, political, and social dimensions; and 8) contribution to international cooperation.

The 2012 draft also stipulates a few basic definitions of cyberspace, cyberattack, cyber security, cyber defense, and critical infrastructure. While the definitions offered for the first four terms are very simple, the last one provides some insight into Turkey's strategy and priorities. Critical infrastructure is defined here according to ten categories that are divided into four levels. The first, core layer includes only the category Informatics. The second layer includes the categories Energy, Finance, and Health; the third consists of Food, Water, and Transportation.

The final layer encompasses Defense (Military), National Security, and CBN capabilities (Chemical, Biological, and Nuclear).

Furthermore, the draft delineates nine strategic objectives that should guide Turkey's efforts to secure the country's cyberspace. Priority is given to, first, defining and mapping critical infrastructure in order to take protective measures depending on the different levels of sensitivity mentioned above; second, creating a culture of cyber security and informing and familiarizing citizens with the latter, with special attention to people in management roles; and third, supporting the development and implementation of Turkish cyber security technologies. Fourth, the draft envisioned the development of legislation related to cyber security and cyber defense that is both deterrent and implementable, in line with international law and regulations. Such legislation would ensure access to information and freedom of expression; protect the privacy of communication, personal information, and personal opinion (except in cases of court order); ensure usability, integrity, privacy, undeniability, and authentication; improve the knowledge of law enforcement on cybersecurity legislation; strengthen security and counter measures against cyber and economic espionage; and create a legal basis for information sharing between state agencies and private companies.

The fifth strategic objective expressed by the 2012 draft focused on investment in human capital and the training of new cyber security experts by creating undergraduate and graduate programs in cyber security; adding compulsory courses on the subject at universities; creating relevant research institutions; giving priority scholarships for cyber security degrees; sending students abroad to study at institutions with a proven record of expertise in this sphere; appointing cyber security or information security managers to all public institution; setting a minimum quota for public servants to be trained in these issues; and continually organizing trainings on new developments in the field.

Sixth, the authors of the draft strategy provide a roadmap for the institutionalization of cyber security measures through the creation of several dedicated institutions. These included a National Cybersecurity Council to ensure the implementation of the *Cyber Security Strategy*; Turkish National Cyber Incident Response Teams which would work in close coordination to deal with cyberattacks; and a National Cyber Threat and Vulnerability Examination Laboratory, which would be an office for the surveillance and registration of national and international cyberattack types, cyber attacker and hacker profiles, cyberattack group structures, attack scenarios, reasons and motivations of attack, attack timing, as well as the development of counter scenarios and strategies. Also, the draft proposes the foundation of a National Cybersecurity Excellence Network under the supervision of the Under Secretariat for Defense Industry, without further clarification of its possible structure or mission statement.

The seventh strategic objective concerns strengthening state-academia-private sector cooperation in the realm of cyber security, through sharing knowledge and best practices, technology, and R&D development; increasing awareness and interest on the topic and providing contracted government job opportunities for experts from the private sector; adding Cyber Security to the 1511 coded list of topics that are set as a priority for scholarships by the Scientific and Technological Research Council of Turkey; adding Cyber Security to the SAN-TEZ program (which is supported by the Ministry of Science, Technology, and Industry and encourages dissertation and thesis-level research on topics related to industry and technology); making research on cyber security compulsory for R&D centers if they want to enjoy state support; and setting a yearly minimum quota for the number of master's and PhD students in the field. Finally, the eighth and ninth strategic objectives relate to encouraging or making mandatory the use of nationally developed security systems and services that are approved by international certification agencies; and fostering international cooperation on cyber security issues on the bilateral and multilateral levels, especially when it comes to the sharing of best practices, intelligence sharing, and cooperation on counter-sabotage.

These listed strategic objectives give insights into the draft authors' understanding of the deficiencies in Turkish cyber security, as well as their priorities in this realm. As suggested before, critical infrastructure is at the forefront of their concerns, depending on the levels of sensitivity outlined above. The other eight listed objectives can be interpreted as a summary of perceived deficiencies and weaknesses and possible objectives to work toward. The objective related to creating awareness on cyber security reveals a recognition of the weakest link in cyber security, that is, systems' users. The emphasis on training people in (government) management roles shows that there is still, in these circles, a lack of knowledge on cyber security and its potential consequences and possibly some resistance on keeping up with the times. Similarly, the highlighting of the need for deterrent legislation is related to for the authors' desire to raise awareness about the importance of cyber security.

Furthermore, the promotion of national technologies can be seen in terms of Turkey's broader economic strategy to develop, use, test, and standardize products in the internal market and promote their export; it also fits within the regime's ambition to lessen dependence on foreign countries in terms of technological products and broaden its own zone of influence by becoming a supplier of these products, particularly those related to the defense industry. This same perceived need to develop technologies domestically also informs the draft's strategic objectives related to human capital and the promotion of innovation and entrepreneurship.

It is worth mentioning that this emphasis on the domestic development of defense- related technologies should be understood in light of the 1974 weapons embargo by the United States on Turkey in the context of Turkey's Cyprus Peace Operation to protect Turkish Cypriots from EOKA's massacres. Indeed, historically, the 1974 crisis acted as a major stimulant for Turkey to develop its national defense industry and create related research facilities. Founded in 1975 as a reaction to the embargo, Aselsan became Turkey's major defense company and training facility for defense and technology experts, working in close cooperation with the military, security forces, and scientific institutions (CNNTurk 2019). One could argue that the authors of the 2012 draft cyber security strategy were also reacting to a perceived crisis (involving increasingly sophisticated cyber attacks against the country), resulting in the similar creation of new domestic capacity to deal with an external challenge.

Finally, the 2012 draft emphasizes that the strategic objectives must be clear, reasonable, and feasible and adapted to Turkey's economic circumstances, technological development, geopolitical positioning, and natural resources in order to be successfully implemented. This may be interpreted as an attempt to create a mindset that accepts that the cyber security strategy should be a step-by-step roadmap driven by demand and limited by existing resources, avoiding the temptation of overambition driven by aspirations that are not matched by realistic capabilities.

Legislation and Adopted Strategy

The cyber security strategy draft discussed in the previous section was soon followed by the adoption of legislation meant to shore up the country's security in this realm. It also resulted in the adoption in 2013 of Turkey's first *Cyber Security Strategy*. Both are discussed in some detail below.

Legislation

The discussed draft resulted in a Council of Ministers Decree (no. 2012/3842), issued on June 11, 2012, and published on October 20, 2012, in State Gazette no. 28447 under the title "Decision Related to the Conduct, Management, and Coordination of National Cyber Security Efforts." This decree regulates the bureaucratic procedures and structure, the responsibilities of the actors involved in cyber security, and the creation and organization of the National Cyber Security Council and working groups on cyber security. On February 19, 2014, a revision was promulgated in State Gazette no. 28918, whereby the authority to determine the members of the National Cyber Security Council was taken from the Ministry of Transportation, Maritime, and Communications and given to the Council of Ministers (under the authority of the Prime Minister); the assignment of duties and working principles of the National Cyber Security Council was relegated to the Prime Minister's authority; and small additional topics such as the promotion of safe use of the Internet were added to the program.

In addition to the Council of Ministers' Decree, the effort to regulate Turkey's cyber security sphere is supported by other existing legislation. This includes first and foremost the no. 5809 Law on Electronic Communication, which regulates the field of electronic communication in sixty-nine articles. Moreover, the Turkish Penal Code also includes several articles related to cyber security that can be found under the section "Crimes against Society Part 10: Crimes Committed in the Field of Information." The specific articles are Article 243 on intrusion into IT systems; Article 244 on the disruption, destruction, and replacement of data on IT systems, as well as the achievement of unlawful benefits through the exploitation of IT systems; and Article 245 on the misuse of debit and credit cards. The Penal Code also includes articles that are not directly classified as relating to cyber incidents but that are applicable whenever there is an important IT related aspect involved in a specific crime. These are mentioned in the section on "Crimes against Individuals," under Article 124 on the disruption of communication; Article 125 on Insulting; Article 132 on privacy of communication; Article 142 on theft; and Article 158 on fraud.

Turkey is also party to the Budapest Convention on Cybercrime of the Council of Europe, which is the only international convention on cyber crime allocating responsibility to the signing parties. The Budapest Convention serves as a guideline for national legislation and strategies and provides a backbone for the Octopus Conference organized by the Council of Europe, where current and developing issues in cyber security are addressed. As a signatory, Turkey is subject to the Convention's protocols on copyright infringements, computer-related fraud, violations of network security, and child pornography. The Convention was signed by Turkey in 2010, ratified in 2014, and entered into force in 2015, fourteen years after the Convention was opened for signature by the Council of Europe in 2001 (Council of Europe 2019). This delay can be explained by the absence of an official Turkish national cyber security strategy until 2013. The Convention also has an additional protocol on criminal acts of racism and xenophobia committed in cyber space, which Turkey signed in 2016 (thirteen years after it was first open for signature, in 2003) but which has not yet been ratified. (Council of Europe 2019).

Strategy

Building on the draft and Decree published in 2012, Turkey's first *National Cyber Security Strategy and Action Plan* was published in January 2013 by the Ministry of Transportation, Maritime, and Communications. It was this specific ministry that was given the role of drafting, managing, updating, and coordinating policies, strategies, and action plans related to cyber security. Although it was made public in

January 2013, the *Strategy* entered into force with a lag of almost six months after its publication in State Gazette no. 28683 on June 20, 2013, probably due to bureaucratic delay.

The *Strategy* resembles closely the draft published in 2012, although it is somewhat more detailed and expanded. First, where the draft defines only four basic terms, the *Strategy* defines twelve of the latter and makes several terminological distinctions, such as between cyber space and cyber structures and between public and private information systems. Furthermore, in addition to national security aspects and the need for the development of expertise, the *Strategy* attaches great importance to the link between cyber security and economic competitiveness, referring explicitly to the dangers posed by economic espionage and sabotage. This is also observable in the definitions given; for example, threats to critical infrastructure now include economic loss in addition to loss of life and threats to national security and public order. Another remarkable addition is the realistic admission that cyberattacks can never be fully exterminated or protected against; accordingly, the goal of the *Strategy* is defined as minimizing the number of cyberattacks and their impact and bringing IT systems to normal in the shortest period of time after an incident.

Similarly, the *Strategy* mentions the need to be realistic when assessing the risks and capabilities related to cyber security, but the subsequent list of risks specific to Turkey is very superficial. It includes: poor national awareness on the topic; poor national coordination between different actors involved in cyber security; a lack of reporting of cyber incidents and attacks due to fear of loss of face and other reasons; cooperation problems caused by deficiencies in national and international legislation; poor user awareness and user errors at the individual and institutional levels; the vulnerability of IT systems in case of natural disasters and the lack of capacity to take the needed precautions; the lack of infrastructure for information management; the lack of knowledge, awareness, and ownership on the issue on the management level; poor institutional infrastructure; lack of personnel; poor levels of expertise and lack of experience among existing personnel; insufficient oversight on cyber security; not taking into account cyber security aspects during the public purchase of products and services; and poor levels of national initiatives on hardware and software development.

In order to overcome these deficiencies and implement the cyber security strategy nationwide, the *Strategy* mandated several strategic actions with an initial deadline of 2014. The first of these concerned the creation of legal regulations and a cyber security terminology glossary, along with structures to support the judicial process and forensics in case of cyber incidents and attacks. The strategic actions also included the establishment of the National Computer Emergency Response Center (USOM, TR-CERC) and sectorial Computer Emergency Response Teams (SOME, CERT) coordinated by TR-CERC; the document also envisioned the strengthening of the national cyber security infrastructure, investing in human capital, and the development of national technologies. The final strategic action mentioned by the *Strategy* involves the extension of the jurisdiction and task range of national security organizations in order to cope with cyber security. Because these strategic actions are very broad, 29 specific missions and 95 sub-missions are defined and delegated to specific ministries or agencies, with explicit deadlines for implementation.

Two years after the *Strategy's* adoption, seven workshops were organized in 2015 to evaluate and revise it. Referred to as the Platform for Common Mind, these meetings involved the participation of 126 experts representing 73 institutions and resulted in the subsequent publication of an updated version of the *Strategy*, which outlined Turkey's priorities in the cyber security realm until 2019. Starting from the introduction, this document stresses again that cyber security is a crucial aspect of Turkey's economy (this time mentioning e-commerce as an area of concern for the first time) and stresses the threats of cyber espionage with a focus on economic espionage, particularly in the defense industry. Additionally, the language of the updated *Strategy* indicates that there was some work on aligning Turkey's cyber security strategy with the EU, OECD, and NATO.

It should be noted that both the 2012 draft and the 2013 *Strategy* stipulated that the strategy would be updated at least once a year. However, in practice, this has not occurred, which probably explains why specific quotas and time limits related to projects and publication of updates were deleted from the updated document. More than that, the latter now includes language specifying that "proposed actions that could not be finalized and goals that could not be accomplished before the publication of the following strategy document would be transferred to the next document." These two changes give the impression that the initial enthusiasm among relevant policy actors has diminished and given way to the acceptance of delay and postponement. A similar conclusion can be reached when we examine the website of the Information and Communication Technologies Authority (Bilgi Teknolojileri Kurumu - BTK), one of the major actors in Turkey's cyber security. As of this writing the webpages related to cyber security were out of date, with the last update being from December 2017. In the aftermath of the July 15, 2016, coup attempt and the subsequent reform of the presidential system, many things have changed in terms of state structure and thus also the structure related to cyber security. However, none of these changes have been reflected via updates to relevant strategy documents and government websites. For example, the BTK website still makes references to the Under Secretariat of Public Order and Security, which was shut down in 2018, and even to the Telecommunication and Communication Authority, which was shut down in 2016, one year before the BTK's webpage was last updated (Bilgi Teknolojileri Kurumu 2017).

Still, two important developments were included in the updated *Strategy* from 2016. The first was the recognition of Internet and social media addiction as a cyber security risk affecting the Turkish population; in response, a special Parliamentary Research Commission was set up to investigate this issue and come up with solutions (Takvim 2019). Second, the updated *Strategy* mentions that, in addition to protective measures, Turkey also aims to build proactive cyber capabilities for preventive and pre-emptive purposes.

Besides the official declarations mentioned above, certain other elements influencing Turkey's cyber security and intelligence strategy are worth mentioning. These include, but are not limited to, far-reaching digital surveillance, the struggle against terrorist propaganda and social engineering, limitations on Internet access, unlawful wiretapping and "tape scandals" that reflect the abuse of power due to lack of oversight, vulnerability against cyber sabotage, lack of data protection, and breaches of privacy.

More specifically, Turkey's surveillance and wiretapping culture has been the object of much criticism. The 2018 brutal killing of journalist Jamal Khashoggi subjected the country's surveillance capabilities and activities to international media scrutiny. Turkey's foremost evidence against Saudi Arabia in this case was a recording of discussions in the Saudi consulate in Istanbul; this led to suspicions that Turkey was eavesdropping on foreign representations inside its borders. Further outcry was caused by public leaks of unlawful secret voice recordings related to national topics (Global News 2018), particularly leaks related to talks between Turkish intelligence and the Kurdish PKK, a high-level security meeting on the Syrian conflict, and alleged corruption cases. In addition, over the past few years Turkey has witnessed dozens of sex-tape leaks and blackmailing of politicians who were forced to resign, including the most powerful opposition figure Deniz Baykal. Before the leak, Baykal accused the director of MIT's Department of Electronic-Technical Intelligence (which is responsible for electronic surveillance) with being an agent of the Gülen sect. This group, which is considered a terrorist organization by the Turkish state, already had the reputation for unlawful eavesdropping on politicians, journalists, academics, businessmen, police and military personnel, judges, prosecutors and numerous others, including foreign targets (CNN Turk 2018).

Structure

The June 12, 2012, Council of Ministers Decree outlined what would henceforth be Turkey's main cyber security structure. The Cyber Security Council under the presidency of the Minister of Transportation, Maritime, and Communications was designated as the core organ responsible for cyber security, with the participation of the Undersecretaries of the Interior, Foreign Affairs, National Defense, Transportation, Maritime, and Communications, the National Intelligence Agency (MIT), the Army General Staff Head of Communication Electronics and Information Systems (MEBS), the Director of the Information and Communication Technologies Authority (BTK), the Director of the Scientific and Technological Research Council

(TÜBITAK), the Director of the Financial Crimes Investigation Board (MASAK), and other officials who may be appointed by the Ministry of Transportation, Maritime, and Communication (this structure was somewhat modified by the subsequent 2014 revision mentioned above). As it is clear from this structure, the Ministry of Transportation, Maritime, and Communications is the main actor in terms of Turkish cyber security and is given the task to draft, manage, update, and coordinate policies, strategies, and action plans to protect the country in this realm. The actual execution of protection and reaction against cyberattacks is carried out by the CERC and CERT teams.

The rest of this section examines more closely the various agencies involved in Turkey's cyber security structure, touching upon their respective roles and internal organization. As we are examining a security-related topic we will also touch upon the role of security and intelligence organizations within the cyber security ecosystem. However, it should be noted that remarkably little has been written about Turkey's four main security establishments, and that this enigmatic character of the Turkish security establishment is no different when it comes to cyber security. Still, as we will see, some of these actors are more transparent than others.

Informational and Communication Technologies Authority (Bilgi Teknolojileri ve Iletisim Kurumu – BTK)

Receiving its tasks from Law no. 2813, Law no 4502 and Law no 5809, The BTK receives its mandate from Laws no. 2813, no. 4502, and Law no. 5809, and is an important actor in terms of regulating, controlling, and auditing the Internet and Internet access, the communication sector, and cyberspace. The institution has considerable power and freedom, defined under Article 4(3) of its management regulations document (Bilgi Teknolojileri ve Iletisim Kurumu 2011). The same document defines the BTK's different duties and jurisdiction under Article 5. Besides the CERC and CERTs, which are discussed below, the departments within BTK dealing with cyber security are the Information Systems Department, Information Technology Department, Internet Department, Department for Access and Tariffs, Sectorial Research and Strategy Development Department, and the Technical Operations Department.

According to publicly available information from the BTK itself, the agency's most important activities related to cyber security are cyber security exercises (Bilgi Teknolojileri ve Iletisim Kurumu 2017). The goals of these exercises are testing, evaluating, and developing counter capabilities; evaluating and improving coordination inside and between the actors involved; and improving awareness on cyber security issues. Since 2015, four exercises have been organized, with the participation of the Ministry of Transportation, Maritime, and Communications, TÜBITAK, and Istanbul Technical University.

Turkey Computer Emergency Response Center (TR-CERC)

The TR-CERC, which is responsible for the dealing with cyber incidents on the national level, is housed within the BTK. It plays the main coordinating role for information related to cyber incidents with different government and private actors and is responsible for alarms, warnings, and announcements on cyber incidents and the detection of attacks. In terms of security updates, the TR-CERC webpage is updated regularly; however, the content is only a copy-paste of updates from a limited number of major companies such as Cisco, Mozilla, Microsoft, Apple, Oracle, Google, and Adobe. When it comes to announcements from the TR-CERC itself, the alert list only contains five alerts published between January 2015 and August 2017, of which four are actually related to the activities of the Computer Emergency Response Teams (see below); the content of each alert is limited to no more than a few lines. Additionally, TR-CERC provides twelve awareness-raising documents on its website. However, these are also outdated, being from between 2014 and 2016. The outdated nature of the website can give a distorted image of TR-CERC's activities, suggesting that it is far less active than is in reality; the issue here is one of a lack of transparency, characteristic generally of the Turkish bureaucratic system.

Computer Emergency Response Teams (CERT)

The details related to the regulations, tasks, and responsibilities of the CERTs are provided in State Gazette No. 28818 (Resmi Gazete 2013). The structure of the CERTs is divided according to sectors and infrastructure, with the teams technically part of the specific institution within they are housed. The CERTs are responsible for taking measures to ensure information security of a specific sector or a particular institution; protecting a specific sector or institution against cyberattacks; taking measures to lower the damage in case of an attack; reacting against possible attacks; ensuring information flow with different partners; and ensuring 24/7 preparedness and availability. The information given on BTK's own website mentions that the CERT on electronic communication within BTK consists of a total of six personnel, including one coordinator and five experts. In a recent interview on CNN Turk, BTK director Omer Fatih Sayan stated that there are around three thousand legally employed cyber security experts working 24/7 at more than one thousand CERTs across Turkey (Sayan 2019).

The following diagram from the BTK website gives a general overview of the relationship and hierarchy between the TR-CERC and CERTs:

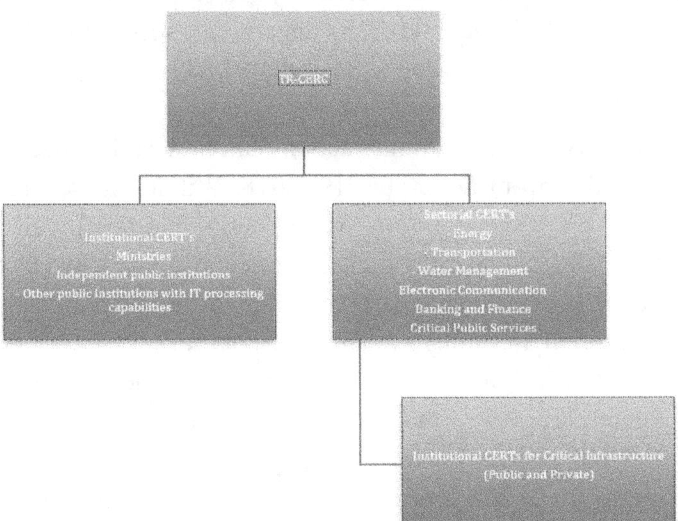

Source: https://www.btk.gov.tr/usom-ve-kurumsal-siber-olaylara-mudahale-ekibi

General Directorate of Security (Emniyet Genel Mudurlugu)

The Turkish police force, officially known as the General Directorate of Security and housed within the Ministry of Interior, has undergone significant evolution in both structure and duties over the years. Prior to 2011, cyber incidents and crimes were dealt with by the police forces in a decentralized way. Efforts that dealt with cyber related aspects of investigations were seen as supportive services. As there was no specific cyber department, relevant personnel working at different departments at the headquarters or local police departments were assigned ad hoc. In 2011, these efforts were consolidated within the Department of Combating Information Technology Crimes under Ministerial Decree 2011/2055. Based on another ministerial decision, on February 28, 2013, the department was renamed the Department of Combating Cyber Crimes. The department's mission statement declares its purpose as "reacting against crimes committed through the use of or targeting of information technology systems, preventing misuse of information technology systems, making assessments related to cybercrime threats, training forensic computing and cybercrime experts and inspectors, working on an international level to combat cybercrimes, and creating public awareness" (Emniyet Genel Mudurlugu sd). The department strives to achieve these objectives with a budget of around 86 million Turkish Lira (Emniyet Genel Mudurlugu 2018).

An overview of this unit's actions can be seen in a declaration on its website dating from December 19, 2018. Within the department, the areas of work are divided among twelve desks that do surveillance and online patrols 24/7, specifically focused on social media; citizens are given the opportunity to assist the department by reporting suspicious webpages or profiles through an online system. The internal division of the twelve desks is listed as follows: 1) fraud related to payment methods; 2) IT systems related crimes; 3) the online sale of narcotics and weapons; 4) religiously motivated terrorism; 5) illegally infiltrated structures into the state apparatus under the control of the Gülen sect, linked to the July 15 coup attempt; 6) separatist terrorism; 7) extremist left-wing terrorism; 8) security; 9) child abuse and prostitution; 10) illegal gambling; 11) administrative investigation; and 12) a category combining uncategorized crimes under the label "other."

It should be noted that the separate category "security" is an interesting combination of various topics, including different activities designated as crimes, such as "insulting Atatürk, insulting statesman, insulting the Turkish nation/people, social engineering, illegal gathering callouts, fake profiles, insulting religious values, organ trade, threatening, inducement to suicide, and violence against animals." In the event, in 2018 the twelve desks investigated approximately 110,000 social media accounts, around 45 000 users were "identified" (with no further information shared as to what this meant), and 7,109 people were arrested for criminal or terrorist activities online.

Since the official start of the Turkish *Cyber Security Strategy* in draft form in 2012, the cybercrime department of the police force has organized yearly Cyber Crime Workshops, of which the fifth was organized between December 10 and 12, 2018, in Ankara, with the participation of Turkish and international experts from law enforcement, academia, and major companies from the finance sector and defense industry. The main topics of interest were social media and discourse analysis, social engineering, banking fraud, business related cyber crimes, cybercrime in court orders, cryptology and law enforcement, Dark Net, financial crimes, public-private partnerships, and the maintenance of a safe Internet for children. This conference series is also an important chain in the cooperation between Turkey and the European Union and Council of Europe, under the iPROCEEDS joint project for cooperation on cybercrime under the Instrument of Pre-Accession (Council of Europe sd). The police force has also launched safe Internet usage awareness campaigns for children, in the form of brochures and posters with cartoons meant to promulgate knowledge about the dangers posed by malicious people online.

Moreover, the police cybercrime department offers courses on cyber security, of which the titles and syllabuses give an idea of priority topics and the importance attached to bringing personnel up to a desired knowledge level. In total twenty courses are listed that add up to 607 hours of theoretical and practical training, including classes on the legal aspects of countering cyber crime, police patrol in cyber space, Deepweb and Darknet, social engineering, fighting online child abuse, technical analysis for different operating systems, retrieving data, cyber and network forensics, electronic evidence collection, counter intelligence, secret investigation techniques, and surveillance and observation (EGM Siber Suclarla Mucadele Daire Baskanligi sd). An additional course of fifty hours under the name "Branch Training" is included on the online list of courses, without further specification.

Besides the Department of Combating Cyber Crimes, the police force has a separate Department of Information Technology. This department, with a budget of around 216 million Turkish Lira, is responsible for the security of the Directorate of Security's internal communication, computer systems, and networks (Emniyet Genel Mudurlugu 2018). The relevant sub-departments are information security; IT network security; systems security; software; database; and IT projects (EGM Bilgi Teknolojileri Daire Baskanligi sd).

National Intelligence Service (Milli Istihbarat Teskilati – MIT)

Similar to the previously mentioned state institutions, the National Intelligence Service MIT (Milli Istihbarat Teskilati) has experienced important transformations in recent years. Besides the changes in the law regulating the intelligence service and the changes in structure, an important step was taken to open the agency to the public and create transparency related to its workings. As part of this transparency campaign, the MIT invited journalists to the MIT Headquarters and briefed them about the structure and working of the service (TRT Haber 2012). In the following year, the MIT's website was renewed, providing basic information on how MIT works (Hurriyet 2013). The website also launched a career webpage, on which (as of this writing) seventeen of nineteen career opportunities were within IT-related departments, varying from jobs for SIGINT analysts to cyber security and Internet technology experts (Milli Istihbarat Teskilati IK sd). Building on this information, it is plausible to assume that these departments are growing or gaining importance for the MIT.

The agency combines all intelligence tasks in one organization, which means that besides the archetypal intelligence duties such as counter intelligence, security and operations departments, the MIT also incorporates the Department of Signal Intelligence and the Department of Electronic-Technical Intelligence (Milli Istihbarat Teskilati sd), which need more specific specialization and for which many countries opt to set up separate independent agencies. This raises questions related to specialization and optimal deployment of staff (in case of rotation), a possible mentality gap between staff (HUMINT <-> SIGINT), and many other human resources and organization related issues.

The Department of Signal Intelligence is responsible for the interception of signals and processing them through the intelligence cycle. Prior to its transfer to the MIT in 2012, this department was under the Army General Staff Electronic Systems Command (Genelkurmay Elektronik Sistemler – GES). In order to compensate for the probable reducing of SIGINT capabilities at air and sea after the department's separation from the Army, MIT procured UAVs and one ship. The agency purchased ANKA-I UAVs produced by TAI (Turkish Aerospace Industries) (Milliyet 2018) but also showed interest in the Global Hawk produced by Aerovironment (USA) and still cooperates with TUBITAK and Turkish defense companies for other projects (Cumhuriyet 2014). Additionally, MIT has four CASA CN235 airplanes transformed in order to meet SIGINT needs and has considerable satellite technology developed by national and international companies. Besides these, MIT has an order standing for a fully equipped SIGINT airplane, about which no recent news is available (Haber7 2015).

MIT's first ship was ordered in 2012 (Boshporus Naval News 2012), produced in cooperation between STM (Savunma Teknolojileri Muhendislik ve Ticaret A.S.) and ASELSAN, and was delivered on February 9, 2019, under the name TCG Ufuk A-591 (Milliyet 2019). The project was developed officially as a "testing and training ship" but Turkish president Erdogan revealed during the ceremony of the delivery of the ship that TCG Ufuk was to be the country's first intelligence ship, meant to be its eyes and ears on the high seas (T.C. Cumhurbaskanligi Savunma Sanayii Baskanligi 2019). This move is a clear message about

Turkey's ambitions in the region, its commitment to protect its interests on the high seas, and its intent to preserve Turkish security beyond Turkey's borders.

While the above developments clearly demonstrate the strengthening of the SIGINT department's capacity, the current dominant view is that the former has also in recent times suffered some damage due to infiltration by members of the Gülen sect (Sabah 2018). Moreover, there was much critique when SIGINT capabilities were taken from the Army and transferred to the MIT, as possible communication failures or delays could cause serious problems for military operations. Some experts (including the former director of GES) claimed that a probable miscommunication or delayed communication between the MIT and the Air Force was the reason for the shooting down of a Turkish jet by Syrian missiles in 2012 and the shooting down of a Russian Sukhoi Su-24 by a Turkish F16 in 2015; allegedly, the intercepted intelligence was delivered to the President before reaching the Army General Staff, resulting in a critical delay (Sozcu 2015; T24 2015). Other observers were convinced that the miscommunication was caused by Gülenist sabotage.

The second MIT department working on the cyber dimension is the Department of Electronic-Technical Intelligence. The duties of this unit are defined as "interception and evaluation of communication in order to detect state secret disclosure/leaks and terrorist activities, [and] keeping records of the intercepted data" (Milli Istihbarat Teskilati sd). Additionally, the unit is responsible for "Imagery Intelligence (IMINT), deciphering intercepted encrypted data, analysis of intercepted voice and imagery, [and] counter activities against cyber threats and attacks." As many state institutions, this department appears to have been the target of infiltration at the highest level by the Gülen sect. In a striking example, the previous director of this department Basri Aktepe has been accused by various opposition actors of being a Gülenist infiltrator for more than ten years (Terkoglu 2011), and is now being prosecuted for allegedly participating in the coup attempt and being a member of a terrorist organization (CNN Turk 2019).

The third MIT department active in cyber security is the Counter-Intelligence Department (CID), with a focus on countering foreign espionage threats. As part of this effort, this unit introduced briefings on counterespionage and protection against cyber threats as far back as 2009, a very proactive effort predating the official start of Turkey's cyber security strategy. Since then, more than 18 thousand people from 390 different institutions have been briefed by CID (SputnikNews 2018). Finally, the MIT in general provides "strategic cooperation" briefings and recently introduced "regional security evaluation" briefings directed toward the academic community (Gazete Yolculuk 2019).

Army General Staff Department of Communication Electronics and Information Systems (Muhabere Elektronik ve Bilgi Sistemleri - MEBS)

The Department of Communication Electronics and Information Systems under the Army General Staff is responsible for the management and protection of military communication systems, electronic warfare, information systems, and IMINT (Kara Kuvvetleri Komutanligi sd). The department is organized according to the Network-Centric Warfare doctrine in order to enhance operational effectiveness. Recently, the brigadier general in charge of this unit was appointed to the NATO Communications and Information Agency based in Brussels (Yeni Akit 2019).

A department with a similar name also exists within the Presidency of Defense Industries of the Presidency of the Republic of Turkey, which oversees and manages the development of new projects and products related to this area (T.C. Cumhurbaskanligi Savunma Sanayii Baskanligi sd). This Presidency also houses a separate department on Cyber Security and Information Systems.

Gendarmerie General Command Branch of Communication Electronics and Information (Jandarma Genel Komutanligi Muharebe Elektronik ve Bilgi Sistemi - MEBS)

The Gendarmerie has its own communication electronics and information branch. The similarity of its name with that of the above-discussed department of the Army General Staff is probably because the Gendarmerie was based within the Army General Staff for a long time, being transferred to the Ministry of Interior only after the 2016 coup attempt. The cyber capabilities and duties of the Gendarmerie are not very clear. However, it is possible to determine some of the unit's activities via its advertised job vacancies (T.C. Icisleri Bakanligi Jandarma Genel Komutanligi 2018). For example, the currently listed vacancies do not provide any information on the job content. However, the education requirements listed (graduate degrees in electronics, computer sciences, communication technology, and Internet and network technology) are suggestive. One of MEBS's probable activities is the management and security of the Gendarmerie Integrated Communication and Information System (Jandarma Entegre Muhabere ve Bilgi Sistemi – JEMUS), which is in fact the integrated working of the Gendarmerie General Command Incidents Information System and the Gendarmerie Integrated Intelligence System.

Scientific and Technological Research Council of Turkey (Turkiye Bilimsel ve Teknolojik Arastirma Kurumu – TUBITAK)

TUBITAK is Turkey's state-run agency responsible for the setting-up, funding, management, coordination, testing and evaluation, and commercialization of scientific projects and technological R&D within state defined targets, in close collaboration with academia, government, private companies, and international institutions. As Turkey's foremost R&D agency, TUBITAK plays an important role in cyber security and the development of new national technologies. Within TUBITAK, the Informatics and Information Security Research Center (Bilisim ve Bilgi Guvenligi Ileri Teknolojileri Arastirma Merkezi – BILGEM) is the lead actor when it comes to cyber and information security. More than 1600 scientists work at different research institutes under BILGEM, including 1) the National Research Institute of Electronics and Cryptology, 2) the Information Technologies Institute, 3) the Advanced Technologies Research Institute, 4) the Cyber Security Institute, and 5) the Software Technologies Research Institute. These institutions provide important solutions, varying from secure communication, electronic intelligence, smart transportation, electronic warfare systems, cloud computing, and disaster management. Critical projects like the new electronic National ID Card, the GÖKTÜRK Reconnaissance and Surveillance Satellite and its Crypto System, the National Military Messaging Handling System, and the MILCEP project for crypto mobile phones are just a few of the many projects carried out under BILGEM's auspices. TUBITAK also plays an important role in the development of new surveillance technologies and cyber security solutions for the National Intelligence Agency in close cooperation with Turkey's leading defense companies, such as ASELSAN and HAVELSAN. For this reason, TUBITAK has also been a target for foreign intelligence services.

Other Agencies

The picture of Turkey's current cyber security structure would be incomplete without at least a brief mention of the Disaster and Emergency Management Authority (AFAD), which includes cyber threat and cyber security in its Disaster Management Glossary. AFAD has a coordination role related to cyber threats and incidents against critical infrastructure, which the agency refers to as "technological disasters." However, AFAD's role is only the coordination of information. Responsibility and execution is delegated to numerous different national and local institutions. For this reason AFAD has been lobbying for a clearly defined structure with clear boundaries related to responsibility, accountability, task division, optimized communication and coordination, and the development of clear critical infrastructure security plans (Afet ve Acil Durum Yonetim Baskanligi 2014).

Finally, the Cyber Security Initiative within the Internet Development Council is a platform where stakeholders gather for sharing information and discussing possible improvements on cyber security, with the results of these discussions being presented to the Ministry of Transportation, Maritime, and Communication (Bilgi Teknolojileri ve Iletisim Kurumu 2017). In addition, this platform also aims to inform and create awareness among citizens and SMEs, conduct sectorial risk analysis, determine cyber security standards, and publish reports and manuals on the topic, among other activities.

History of Cyber Attacks and Incidents in Turkey: The Case of Hacktivism

Having presented Turkey's *Cyber Security Strategy* and the institutions tasked with achieving its goals, we conclude this article with a discussion of one of the major cyber security threats faced by the country in the twenty-first century, namely, hacktivism. Doing so highlights the importance of Turkey's efforts in this realm and raises questions about the success of the present *Strategy* and its implementation.

In the twenty-first century, Turkey has witnessed cyber incidents and attacks on an unprecedented scale, with devastating results, a phenomenon that makes it difficult to understand the delay in state action to develop a cyber security strategy. The incidents that are available through open sources have mainly involved cyberattacks for economic purposes, sabotage, hacktivism, and crimes between citizens for various reasons. Besides these, there have undoubtedly also been incidents related to espionage that have not been made public.

Acknowledging that cyber incidents with a fraudulent economic aim are a serious problem for Turkey, such as in the example of the hacking of HSBC Turkey where the account and credit card information of 2.7 million customers was stolen, this article will only consider cyber incidents with a political dimension, known more specifically as hacktivism, which has been a serious challenge for Turkey. Numerous incidents have taken place in recent years, with the main actors being Turkish groups sometimes operating with the support of foreign or international hacker communities. Hacktivist communities in Turkey can be divided in two clearly separate groups: the nationalist hacktivist groups who perform cyber attacks in line with what they see as the interest of the Turkish state and nation, and secondly the left-wing hacktivist groups who perform cyber attacks in line with their political beliefs and self-proclaimed responsibilities as cyberguerillas. Finally, it should be noted that while hacktivism incidents have in general demonstrated the vulnerability of the Turkish state institutions to cyberattacks, some hacker groups have stated explicitly that their actions are meant to create awareness of this problem and to force state institutions to take up the responsibility to protect their systems (Yeni Mesaj 2006).

Over the past several years, the webpages and information systems (such as PolNet) of Turkish security forces have been hacked repeatedly, with the retrieved information being leaked to the public. In the leaks after the hack of the Ankara Security Directorate (police force), the hackers also publicized that the access code for the Directorate's database was simply "123456" (Posta 2012). It should be noted that some experts claim that the consequences of the hacks of the Security Directorate were more severe than was made public.

In the meantime, the Army has also been a regular target of cyberattacks. For example, the Land Forces Command was hacked several times in 2012, and some details of personnel were made public. The hackers left the following message: "If we can get access to this information, imagine what foreign intelligence services could do," attracting attention to the deficiencies in Turkey's cyber security (Sabah 2012). Cyberattacks against the Army have taken very creative forms. On March 1, 2019, the Army General Staff sent out an alert to its personnel related to online games that were used for cyberattacks, wherein military personnel were targets of social engineering to retrieve intelligence (Internethaber 2019). Similarly, the MIT has suffered attacks by hacktivists on several occasions in the form of DDoS attacks and hacks to the MIT website.

Other governmental institutions, mostly related to security or strategic projects, have been regular targets. The ministries of urban planning and energy have been repeatedly targeted by self-proclaimed environmentalist hackers opposing hydroelectric power plants and projects for nuclear power plants with slogans such as "Nuclear power plant is fascism." Major power outages on national scale have also been attributed to cyberattacks. For example, the unprecedented power outage that occurred on March 31, 2015, and stopped life in Turkey for twelve hours (affecting hospitals, airports, traffic, and infrastructure such as water, but also simple things like elevators and refrigerators) was ascribed by some sources based on leaked documents to hacking by Iranian hackers (T24 2017). The Turkish Ministry of Energy also attributed power outages that happened between December 2016 and January 2017 to cyberattacks and sabotage attempts from US-based platforms and claimed that subsequent infiltration attempts were countered (Sputnik 2017). In these cases, the question of whether the attacks were carried out by hacktivists (with or without government support) or were in fact the products of foreign government action is still unanswered; they are included here as examples of the scale of the problem Turkey faces.

The Turkish Telecommunication Authority (Telekomunikason Iletisim Baskanligi – TIB), which was transferred to the BTK after the 2016 coup attempt, was hacked by a Marxist-socialist hacker group and the retrieved institutional information was made public. The hacktivists proclaimed their purpose as "protesting the heavy Internet censorship imposed by the TIB," and left an explicit message to the TIB: "You thought you calculated everything, but you forgot the main coordinator of everything. The prohibitor gets prohibited" (CNN Turk 2014). Other hacktivists have performed cyberattacks to protest high Internet prices and low connection speeds or even a high heating bill. Discontent is also visible in the form of targeting political parties and politicians. The website of the ruling Justice and Development Party is attacked regularly, in the form of deceptive messages, mostly with communist imagery. On what could be considered a positive side, Turkish hacktivists also perform attacks against child abuse networks by taking down their networks.

Both left-wing and right-wing hacktivists also perform attacks against foreign state targets, which are ironically most of the time Turkey's close allies. Numerous targets (including police forces, the ministries of defense and newspapers) abroad have been subjected to attacks for reasons ranging from protests against racist policies against minorities, active or passive state support for anti-Turkish terrorist organizations, and anti-Turkish sentiment in politics and media (Nieuwsbald 2007; Yeni Akit 2017; Knack 2019). For example, after a spate of cyberattacks in March 2013 against Israeli targets to protest Israel's policies towards Palestinians, Turkish hackers announced the following: "The real large attack will be on the 7th of April on a global scale, we will erase Israel from the Internet" (Oda TV 2013). Interestingly, protective measures against the subsequent cyberattacks in April failed despite the fact that these attacks were announced weeks before (Oda TV 2013). Similarly, American state institutions have also been exposed to Turkey- based attacks (Hurriyet 2007; Mynet 2010; Yeni Akit 2018; Sabah 2016), as have European countries.

Finally, multinational organizations have been targeted, often around issues of deep concern to Turkish society broadly speaking. For example, the website of the United Nations was hacked in 2007 and 2014 in order to protest the policies of Israel and the United States toward Palestine (Hurriyet 2007) (Milliyet 2014). Another attack against the UN was carried out to protest the ethnic persecution by China of Uyghur Turks (HaberTurk 2009).

Conclusion

As mentioned in the introduction to this article, Turkey began its cyber security campaign with high ambitions and enthusiasm. This was largely fuelled by the country's need for security, made obvious by a series of related incidents, but also by its desire to become a global player in all possible fields. When we examine the *Cyber Security Strategy* adopted in 2013 (Turkey's first official document of this nature), both the failure to carry out the mandated updates and lacklustre implementation suggest that this early enthusiasm did not last long. At the same time, this impression is strengthened by the lack of governmental transparency in this sphere, or at least in the reluctance of the state agencies to share information.

Moreover, based on the fact that cyberattacks, crimes and other incidents (including but not limited to the hacktivism discussed in the above section) have continued at a high pace since the *Strategy*'s adoption, it is unlikely that the policy's main declared goals are being reached, particularly when it comes to the protection of critical infrastructure and individual and institutional data. Further research is needed to answer the question whether this is due to the failure or shortcomings of the Turkish strategy and agencies, or whether it is due to an increase in the number and sophistication of attacks and attackers.

More importantly, when looking at the practice, it is unclear whether any of the explicitly listed values meant to inform the cyber security strategy are truly being taken into account. Notably, critics have pointed to censorship and lack of oversight resulting in a culture of unlawful misuse of cyber capabilities, particularly visible in the form of the "tape scandals" that have emerged over the past several years. This situation suggests the need to open a discussion on what is meant by (cyber)"security," its depth and scope, and the importance of democratic oversight of cyber security capabilities.

At the same time, on the positive side, the 2013 *Strategy*, related efforts, and legislation demonstrate that Turkey attaches great importance to complying with international (technological) standards (mainly due to economic reasons), as well as to cooperation with NATO, EU, OECD and its allies. It is important that these commitments continue despite attempts to subvert Turkish public opinion and sabotage attempts by domestic and foreign actors.

Moreover, an evaluation of the cyber security structures created in the aftermath of the *Strategy*'s adoption shows that the proposed agencies and institutions were created and activated inside the mandated time limits. The existence and apparent potential of the created agencies also suggests that Turkey has the intention to achieve serious preventive and pre-emptive capabilities. However, the requisite human capital and expertise have seemingly not yet been arrived at. This conclusion is partially due to the fact that the results of investments in human capital are only visible after a period of time (meaning that it may be too early to draw either negative or positive conclusions in this regard), as well as to the loss of capacity and expertise stemming from the political turmoil of recent years. Political turmoil and battle for power and jurisdiction inside Turkey's security establishment resulted in the restructuring of the still young cyber security structures, which in turn caused critique related to incidents caused by communication failures.

In addition to political turmoil, the Turkish ruling elite is aware of the vulnerability of the Turkish population to cyber and social media addiction, which in combination with a lack of awareness on safe use of the Internet constitutes a significant challenge. Together, political turmoil and social vulnerability form a serious problem, as the existing tensions are exploited for social engineering purposes by both domestic and foreign actors. Another major issue relates to the leaks of information retrieved through hacking, more specifically hacktivism. In order to overcome these issues, Turkey has visibly tightened the grip on Internet access and increased its digital surveillance; however, critics have suggested that there may be political motives behind these censorship efforts that trump purely cyber security concerns.

When looking at the final picture, Turkey started dealing with its existing cyber security problems with a significant delay but managed to set up a reactive system in a considerably short period. In order to reach the defined goals within the boundaries of the explicitly stated values while finding a healthy equilibrium in the security-freedom nexus for this new state structure, the notions of transparency, critique, and discussion should be accepted as fundamental. For the rest, the system must be given time to develop to the intended level and prove itself. Finally, but crucially, democratic oversight of this system must be guaranteed in order to audit its effectiveness and safeguard the fundamental values of liberty and justice.

Bibliography

Afet ve Acil Durum Yonetim Baskanligi. *Kritik Altyapilarin Korunmasi Yol Haritasi Belgesi 2014-2023.* Ankara: AFAD, 2014.

Bilgi Teknolojileri ve Iletisim Kurumu. *Siber Güvenlik İnisiyatifi*. December 15, 2017. https: //www.btk.gov.tr / siber-guvenlik - inisiyatifi (accessed February 28, 2019).

Bilgi Teknolojileri ve Iletisim Kurumu. *Siber Guvenlik Kurulu.* December 15, 2017. https://www.btk.gov.tr/siber-guvenlik-kurulu (accessed February 24, 2019).

Bilgi Teknolojileri ve Iletisim Kurumu. *Siber Guvenlik Tatbikatlari.* December 15, 2017. https://www.btk.gov.tr/siber-guvenlik-tatbikatlari (accessed February 21, 2019).

Bilgi Teknolojileri ve Iletisim Kurumu. *Bilgi Teknolojileri ve Iletisim Kurumu Teskilat Yonetmeligi.* 2011.

Boshporus Naval News. *Turkish National Intelligence Organisation Asks for A SIGINT Ship.* October 21, 2012. https://turkishnavy.net/2012/10/21/ turkish-national- intelligence – organization – asks – for - a-sigint - ship/ (accessed February 25, 2019).

Boshporus Naval News. *What Does the Exercise Mavi Vatan Mean?* February 27, 2019. https://turkishnavy.net/2019/02/27/what-does-the-exercise-mavi-vatan-mean/ (accessed February 27, 2019).

CNN Turk. *İstihbaratçı polisten Deniz Baykal itirafı*. December 11, 2018. https://www.cnnturk.com/turkiye/istihbaratci-polisten-deniz-baykal-itirafi (accessed March 06, 2019).

CNN Turk. *Eski istihbaratçı Aktepe için FETÖ'den 15 yıl istendi* . February 7, 2019. https: //www.cnnturk.com / turkiye / eski-istihbaratci-aktepe-icin-fetoden-15-yil-istendi (accessed February 27, 2019).

CNN Turk. *Redhack TIB'i "hack"ledi.* March 28 2014. https://www.cnnturk.com/haber/bilim – teknoloji / internet/redhack-tibi-hackledigini-duyurdu (accessed February 28, 2019).

CNN Turk. *ASELSAN'dan 2 milyar liralik ciro.* May 9, 2019. https://www.cnnturk.com/ekonomi/aselsandan-2-milyar-liralik-ciro (accessed May 17, 2019).

Council of Europe. *Chart of Signatures and Ratifications of Treaty 185.* May 18, 2019. https:// www.coe.int/en / web / conventions/full-list/-/conventions/treaty/185/signatures?p_auth=srkvCSDR (accessed May 18, 2019).

Council of Europe. *Chart of Signatures and Ratifications of Treaty 189 Additional Protocol.* 18 May 2019. https://www.coe.int/en/web/conventions/full-list/- /conventions/treaty/189/signatures?p_auth=srkvCSDR (accessed May 18, 2019).

Council of Europe. *iPROCEEDS – Targeting Crime Proceeds on the Internet in South Eastern Europe and Turkey.* https://www.coe.int/en/web/cybercrime/iproceeds (accessed March 04, 2019).

Cumhuriyet. *MIT kitalararasi istihbarat toplayacak.* September 1, 2014. http://www.cumhuriyet.com.tr/haber/turkiye/113117/MiT_kitalararasi_istihbarat_topl ayacak.html (accessed February 25, 2019).

EGM Bilgi Teknolojileri Daire Baskanligi. *Bilgi Teknolojileri Daire Baskanligi.* http://www.bilgiteknolojileri.pol.tr/Sayfalar/default.aspx (accessed February 24, 2019).

EGM Siber Suclarla Mucadele Daire Baskanligi. "Ulusal Egitimler ve Ders Icerikleri." Bilisim Suclari Akademisi Sube Mudurlugu, EGM Siber Suclarla Mucadele Daire Baskanligi.

Emniyet Genel Mudurlugu. *2018 Yili Performans Programi.* Ankara: Emniyet Genel Mudurlugu, 2018.

Emniyet Genel Mudurlugu. *Basin Aciklamasi.* February 24, 2016. https://www.egm.gov.tr / Duyurular / Sayfalar / Basin – Aciklamasi - 24-02-2016.aspx (accessed February 24, 2019).

Emniyet Genel Mudurlugu. *Siber Suclarla Mucadele Daire Baskanligi.* http://www.siber.pol.tr/Sayfalar/hakkimizda.aspx (accessed February 24, 2019).

Gazete Yolculuk. *MIT ve YOK'ten ortak konferans.* February 27, 2019. https://gazeteyolculuk.net/mit-ve-yokten-ortak-konferans (accessed February 27, 2019).

Global News. *Khashoggi Case Sheds Light on Turkey's History of Spying and Surveillance.* October 22, 2018. https://globalnews.ca/news/4581250/jamal-khashoggi-turkey- surveillance/ (accessed March 06, 2019).

Haber7. *Milli İstihbarat Teşkilatı yeniden yapılanıyor* . December 15, 2015. http://www.haber7.com/ic-politika / haber / 1703695-milli-istihbarat-teskilati-yeniden- yapilaniyor (accessed February 27, 2019).

HaberTurk. *Turkler BM'yi 'hack'ledi.* July 27, 2009. https://www.haberturk.com /dunya / haber / 161133-turkler-bmyi-hackledi# (accessed March 01, 2019).

Hurriyet. *MIT'in internet sayfasi yenilendi.* January 7, 2013. http://www.hurriyet. com.tr / gundem / mitin-internet-sayfasi-yenilendi-22309701 (accessed February 25, 2019).

Hurriyet. *Türk korsanlar BM sitesini 'hack'ledi* . August 12, 2007. http://www.hurriyet.com.tr/dunya/turk-korsanlar-bm-sitesini-hack-ledi-7074666 (accessed May 18, 2019).

Hurriyet. *Turkler Pentagon'u hackledi.* May 14, 2007. http://www.hurriyet.com .tr/gundem/turkler-pentagon-u-hackledi-6515485 (accessed March 01, 2019).

Internethaber. *Turk Silahli Kuvvetleri"nden askerlere PUBG uyarisi.* March 1, 2019. https: //www.internethaber.com / turk-silahli – kuvvetlerinden - askerlere-pubg-uyarisi- 2004466h.htm (accessed March 1, 2019).

Kara Kuvvetleri Komutanligi. *Muhabere Elektronik ve Bilgi Sistemleri.* http://www.kkk.tsk.tr/Siniflar/Mebs.aspx (accessed February 28, 2019).

Knack. *Brussels Hof van Beroep: 'PKK is geen terreurgroep'* . March 9, 2019. https://www.knack.be/ nieuws / belgie / brussels-hof-van-beroep-pkk-is-geen- terreurgroep/article-news-1438437.html (accessed March 9, 2019).

MDAA. *Greeece's Capabilities.* June 25, 2018. http://missiledefenseadvocacy. org/intl- cooperation/greece/ (accessed May 17, 2019).

Milli Istihbarat Teskilati IK. *MIT Kariyer.* http://www.mit.gov.tr/iksayfasi/index. html (accessed February 27, 2019).

Milli Istihbarat Teskilati. *MIT Baskanligi Teskilat Yapilanmasi.* https://www.mit.gov.tr/teskilat.html (accessed February 24, 2019).

Milli ve yerli test ve egitim (istihbarat) gemisi UFUK (A-591) denize inis toreni. 2019. Milliyet. *MIT'in havadaki kulagi ANKA-I.* 27 March 2018. http:// www.milliyet.com.tr/mit-in- havadaki-kulagi-anka-i-gundem-2635272/ (accessed February 25, 2019).

Milliyet. *Türk hacker, facebook Google ve Apple'ı devirdi* . March 19, 2014. http://www.milliyet.com.tr / turk-hacker - facebook-google-ve / gundem/ detay/1853869/default.htm (accessed March 1, 2019).

Milliyet. *Türk korsanlar BM'yi 'hack'ledi.* July 25, 2014. http://www. milliyet.com.tr / turk- korsanlar-bm-yi-hack-ledi /dunya/detay/1916807/ default.htm (accessed May 18, 2019).

Milliyet. *Turkiye'nin ilk istihbarat gemisi denize indi.* February 9, 2019. http://www.milliyet.com.tr/son-dakika-turkiye-nin-ilk-istih-siyaset-2824996/ (accessed February 25, 2019).

Mynet. *Türk hackerlar Pentagon'u hackledi.* March 11, 2010. https://www. mynet.com/turk- hackerlar – pentagonu – hackledi - 110100499780 (accessed March 1, 2019).

Nieuwsblad. *Turkse nationalisten hacken website FOD Defensie.* January 14, 2007. https:// www.nieuwsblad.be / cnt / dmf14012007_057 (accessed March 1, 2019).

Oda TV. *7 nisan'da Israil internetten silinecek.* March 26, 2013. https://odatv.com/7-nisanda- israil-internetten-silinecek--2603131200.html (accessed March 1, 2019).

Posta. *Hacklenen Ankara Emniyeti'nin sifresini acikladilar.* March 6, 2012. https: //www.posta.com.tr/hacklenen-ankara-emniyetinin-sifresini-acikladilar-111978 (accessed February 28, 2019).

Resmi Gazete. "Siber olaylara mudahale ekiplerinin kurulus, gorev ve calismalarina dair usul ve esaslar hakkinda teblig." 2013.

Sabah. *MIT'teki kripto FETO'culere operasyon.* September 19, 2018. https:// www.sabah.com.tr / gundem / 2018 / 09 / 19 / mitteki – kripto-fetoculere-operasyon (accessed February 27, 2019).

Sabah. *Türk hackerlar NASA'yı hackledi!.* February 15, 2016. https://www.sabah.com.tr / gundem / 2016/ 02 / 15/turk-hackerlar-nasayi-hackledi (accessed March 1, 2019).

Sabah. *TSK'ya siber saldırı!.* June 12, 2012. https://www.sabah.com.tr/gundem/2012/06/12/tskya-siber-saldiri (accessed March 01, 2019).

Sayan, Omer Fatih, interview by Cuneyt Ozdemir. *Dr. Omer Fatih Sayan CNN Turk - 5N1K* (February 23, 2019).

Sol Haber. *Anyonymous ve Redhack Israil'i cokertti.* April 7, 2013. http://haber.sol.org.tr / devlet-ve-siyaset / anonymous - ve - redhack-israili-cokertti-haberi- 71035 (accessed March 1, 2019).

Sozcu. *GES'i SIB yaparsaniz...*November 29, 2015. https://www.sozcu.com. tr/ 2015 / yazarlar / saygi-ozturk / gesi-sib-yaparsaniz-997401 / (accessed February 27, 2019).

Sputnik. *Bakan Albayrak'ın siber saldırı açıklamasına, ABD Büyükelçiliği'nden yanıt* . January 10, 2017. https://tr.sputniknews.com/abd/201701 101026708551-bakan- albayrak-siber-saldiri-buyukelcilik/ (accessed March 6, 2019).

Sputniknews. *ABD'deki Yunan lobisi: Trump'ın Türkiye'ye F-35 ambargosundan mutluyuz.* August 14, 2018. https://tr.sputniknews.com/abd/2018081410 3 4746651-helen- amerikan-liderlik-konseyi-halc-turkiye-ambargosundan -mutlu/ (accessed May 17, 2019).

SputnikNews. *MIT'ten 18bin kisiye 'casusluga onlem' egitimi.* December 12, 2018. https://tr.sputniknews.com/turkiye/201812121036586422-mit-binlerce-kisi-casusluga- onlem/ (accessed February 27, 2019).

T.C. Cumhurbaskanligi Savunma Sanayii Baskanligi. *Organizasyon Semasi.* https://www.ssb.gov.tr/Website/ContentList.aspx?PageID=42 (accessed February 29, 2019).

T.C. Cumhurbaskanligi Savunma Sanayii Baskanligi. *Test ve egitim gemisi TCG Ufuk denize indirildi.* February 9, 2019. https://www.ssb.gov.tr/Website/ContentList.aspx?PageID=1681 (accessed February 25, 2019).

T.C. Icisleri Bakanligi Jandarma Genel Komutanligi. *Sozlesmeli Uzman Erbas Basvuru Kilavuzu (2018).* 2018. https://vatandas.jandarma.gov.tr/PTS_ Aday/ personel1/uzman_erbas/Ek_2_Basvuru_ Kılavuzu.pdf (accessed February 28, 2019).

T.C. Milli Güvenlik Kurulu Genel Sekreterligi. *27 Ekim 2010 Tarihli Toplanti.* October 27, 2010. https://www.mgk.gov.tr/index.php/27-ekim-2010-tarihli-toplanti (accessed February 21, 2019).

T24. *2 yıl önce Türkiye genelinde kesilen elektriğin nedeni İran'ın siber saldırısı mı?* July 14, 2017. https://t24.com.tr/haber/2-yil-once-turkiye-genelinde -kesilen - elektrigin - nedeni – iranin-siber-saldirisi-mi,414375 (accessed March 6, 2019).

T24. *Eski MSB Genel Sekreteri: Elektronik sistemler MİT'e devredilmeseydi, Rus uçağı düşürülmeden engellenebilirdi* . November 30, 2015. https:// t24.com.tr/ haber/ eski- msb-genel-sekreteri-elektronik-sistemler-mite-devredilmeseydi - rus-ucagi – dusurulmeden - engellenebilirdi,318548 (accessed March 5, 2019).

T24. *Microsoft'u Türk hackledi* . December 10, 2008. https://t24.com.tr/haber/ microsoftu- turk-hackledi,20168 (accessed March 1, 2019).

Takvim. *Teknoloji bağımlılığı ile ilgili Meclis Araştırma Komisyonu kuruldu* . February 21, 2019. https://www.takvim.com.tr/guncel/2019/02/21/ teknoloji – bagimliligi – ile – ilgili- meclis-arastirma-komisyonu-kuruldu (accessed March 6, 2019).

Terkoglu, Baris. *MİT'İN YENİ DAİRE BAŞKANI FBI'DA EĞİTİLDİ* . October 8, 2011. https://odatv.com/mitin-yeni-daire-baskani-fbida-egitildi-081011 1200.html (accessed February 27, 2019).

TRT Haber. *MIT-Medya Bulusmasinda Neler Konusuldu.* January 5, 2012. https: //www.trthaber.com / haber / gundem / mit-medya-bulusmasinda-neler-konusuldu- 22952.html (accessed February 25, 2019).

TUBITAK. *BİLGEM Informatics and Information Security Research Center-.* http:// bilgem.tubitak.gov.tr / en / kurumsal / bilgem-informatics-and-information- security-research-center (accessed March 5, 2019).

veTeknoloji. *Siber güvenlik strateji çalıştayı yapıldı.* June 20, 2012. https:// www.veteknoloji.net/haber/siber-guvenlik-strateji-calistayi-yapildi-541 37.html (accessed February 21, 2019).

Yeni Akit. *General atamasi Resmi Gazete'de yayimlandi.* February 22, 2019. https:// www.yeniakit.com.tr / haber / general-atamasi-resmi-gazetede-yayimlandi- 623845.html (accessed February 28, 2019).

Yeni Akit. *Türk hacker ABD istihbaratına sızdı! Şok eden 'Afrin' mesajı.* March 4, 2018. https:// www.yeniakit.com.tr / haber / turk-hacker – abd - istihbaratina-sizdi-sok-eden-afrin- mesaji-431233.html (accessed March 1, 2019).

Yeni Akit. *Türk hackerlardan Belçika medyasına soğuk duş.* October 25, 2017. https://www.yeniakit.com.tr / haber / turk-hackerlardan – belcika - medyasina-soguk-dus- 388736.html (accessed March 1, 2019).

Yeni Mesaj. *Ankara Valiliği "hacklendi"*. August 7, 2006. http://www.yenimesaj.com.tr/ankara-valiligi-hacklendi-H1147706.htm (accessed March 11, 2019).

The Use of Cyber Activities as a Weapon: The Empirical Case of Ukraine

Christine Dugoin-Clement (Paris 1 Pantheon Sorbonne and Centre de recherche écoles de Saint-Cyr Coëtquidan)

Abstract

Since the beginning of the war on its territory in 2014, Ukraine has experienced several waves of cyberattacks. Some have been conducted on the Internet, either through trolls and Twitter bots or via direct cyberattacks targeting critical systems. However, other techniques, such as "psyops" using cyber, have also been tested against Ukraine. In many documented instances, such operations have decreased trust in Ukrainian state institutions as well as public organizations and the military hierarchy. The present paper is based on field interviews in 2017 of military personnel and examines how, by means of cognitive dissonance methods, Internet-based attacks may be used to affect trained targets' behavior by increasing their level of uncertainty. Causing a decrease in specific targets' trust in their institutions or hierarchy enables their behavior to be changed.

The article addresses how these objectives are achieved. It also analyses the results of such attacks on a target-especially in terms of levels of trust-and how they deteriorate the target's cognition. Further, it examines how soldiers perceive being a target of Internet-based attacks using cognitive dissonance.

Introduction

Since the Maidan revolution in late 2013, Ukraine has faced war on its eastern territories in Donbass. This ongoing war is not only "kinetic" but has also included a large cyber warfare component. According to President Poroshenko, Ukraine faced 6,500 cyber- attacks on 36 Ukrainian targets between November and December 2017 alone. The attacks included classical intrusions on private and professional emails of persons of interest, as well as Distributed Denial of Service attacks (DDoS) of websites (including government). Larger- scale operations were also conducted, such as an attack on the electronic system used for the presidential election (May 2014) or power cuts affecting 225,000 people in 2015 and the Kiev blackout of December 2016.

Numerous operations targeting human psychology were conducted against Ukraine, notably on soldiers. Some of these operations used multifaceted tools to interfere with the robust cognition acquired by soldiers during training and to weaken trust in their hierarchy and government. The aim was to influence the subjects' behavior unbeknownst to them, a key issue this article will address.

In 2017, we conducted field interviews with active soldiers and veterans. The aim was to conduct a qualitative analysis in the light of cognitive dissonance theories (self-consistency and SSM model), a decision driven by the nature of the field and the specific profile of the subjects studied. Based on the 2017 interviews, we considered the question of whether Ukrainian soldiers' trust had been affected by cognitive dissonance deployed in Internet- based operations. In other words: had their perceptions changed and uncertainty increased enough due to Internet consumption to modify their behaviors? Another key question was to define whether they were aware of being potential targets or not.

Part I of the paper reviews the literature and defines some of the terminology, such as psyops, etc., and explains the theories used in the analysis, such as cognitive dissonance, and trust in military affairs and in the Internet. Part II presents the Ukrainian case study, its methodology, the data collected, and our analysis.

Literature Review and Definition of Terms

Terms such as "cyber warfare" and "psyops" are often loosely defined due to their cross-disciplinary nature. As such, before presenting the theoretical framework we start the paper by defining relevant terms. This will help shed light on the data collected and analyzed, to be presented in Part II.

Clarifying the Terms

For Schaap, "cyber warfare" is "the use of network-based capabilities of one state to disrupt, deny, degrade, manipulate, or destroy information resident in computers and networks themselves, of another state" (Schaap 2009). Other definitions invoke "actions by a nation-state to penetrate another nation's computers or networks for the purposes of causing damage or disruption" (Clarke 2010). Still others divide cyber warfare into two types: strategic and operational. Strategic is explained as "a campaign of cyberattacks one entity carries out on another," while operational "involves the use of cyberattacks on the other side's military in the context of a physical war" (Libicki 2009). For the purpose of this paper, we use Schaap's definition of cyber warfare, as Libicki's refers to physical war by an identified adversary and as such is not applicable here.

In our case, although suspicions fall on Russia, final attribution of cyber operations is not absolutely defined. And while Clarke's definition involves a "nation-state," Ukraine faces both a separatist movement and cyberspace struggle against structures that are not all officially linked to a nation-state. Furthermore, the present paper addresses cyber operations having psychological effects on military staff rather than effects on computers or networks. Accordingly, we use the term "cyber operation" instead of "cyberwarfare." Moreover, our choice accords with NATO's Cooperative Cyber Defence Centre of Excellence (CCDCOE).

This seems wise, given that the Ukraine conflict is linked to Russia's geopolitical strategy. The CCDCOE's definition (found in the Tallin Manual on the International Law Applicable to Cyber Warfare) denotes a broad approach and is defined as "the employment of cyber capabilities with the primary purpose of achieving objectives in or by the use of cyberspace." Moreover, as the Tallinn manual is the result of a transnational study group, it makes sense to use its terminology.[1] Applied

[1] The Tallinn manual (*Tallinn Manual on the International Law Applicable to Cyber Warfare*, published in 2013 and followed by an updated version in 2017) was written on the invitation of the CCDCOE by a transnational group of IT experts and legal scholars. It deals with international law applications in cyber operations conducted by and directed against states, in cyber conflict and cyber warfare; as such, it had to define the terms covered by laws in this field.

to the present subject, the CCDCOE's definition is large enough to embrace cyber psyops in a context of war, which matches our subject perfectly.

There is agreement among experts today that Ukraine was-and is-a test lab for cyber warfare operations and "psyops," or psychological operations (Greenberg 2017). When talking about psyops, we refer to operations aimed at influencing an enemy's state of mind through non-kinetic means. For this purpose, psyops can use chosen information or indicators to influence people's emotions, thoughts, and reactions. Targets will consequently change their minds and behaviors and, as fallout, the behavior of governments, organizations, or groups can also be changed. The word "psyops" first appeared in 1965 although the expression "Military Information Support Operations" (MISO) may also be used. On the Ukrainian information front, when addressing psyops, various channels were used to match the psychological profiles of targets, and the military was no exception. Concerning soldiers, stealth operations were conducted through the Internet, particularly social networks and chat forums. These operations targeted active fighters as well as veterans and used peer group trust to generate doubt and uncertainty toward the government and military hierarchy.

Existing Literature Review

By its very nature, the subject of the paper is at the crossroads of various theoretical frameworks. Due to their make-up, the operations studied require three different approaches. The methodology is based on cognitive dissonance and heuristic biases; the effects observed involved trust in military teams and military psychology; and the way messages are disseminated are essentially Internet-based, requiring channel-based trust.

Cognitive Dissonance and Decision Making

Although cyber operations use new technology, the methods of influence, such as persuasion or psychological destabilization, have been used before. Cyber tools, with their speed and opacity, are a way to designate a strategy, here based on cognitive dissonance. Developed in the 1950s by Léon Festinger, cognitive dissonance theory postulates that when two cognitions are opposed (said to be "irrelevant") in someone's mind, he or she will experience a "motivational psychological discomfort." The discomfort is "motivational" because in order to feel comfortable again, the subject will behave in such a way as to reinforce or weaken one of these cognitions. Aronson (1968, 1992) improved this theory by casting it in a more functional light. He developed the "self-consistency" variation, in which dissonance arises in people from contradictions between individual behavior/action and their perception of themselves. In this theory, the more people's perception of themselves is positive, or whose values are socially considered as positive, the more they will feel cognitive dissonance. And individuals whose lives are at risk are also more sensitive to cognitive dissonance (Jonas, Greenberg, & Frey 2003).

It seems clear that subjects such as soldiers match these prerequisites perfectly. Effectively, soldiers are associated with values perceived as highly positive, such as sacrifice for the greater good (related to heroic mythology), regularly putting their lives at risk, and facing extreme situations (that is, situations that are unpredictable, risky, and subject to rapid change). Further, empirical experience strengthens cognition. Since it helps soldiers stay alive, training is validated by field experience and external operations (EOs). Training is therefore internalized, and soldiers follow it completely, consistent with Festinger (1956), for whom a cognition can be the result of past experience. In the case of soldiers, a cognition is initially learned during training, then later reinforced through validation by everyday life experience in extreme situations. According to theory, the more the cognition linked with their values is strengthened, the greater their reaction toward inappropriate behavior will be, functioning as an emotional reaction by-passing the thinking filter: more a reaction than a reflection.

Another relevant development of the theory is the Self Standard Model (SSM) of Stone and Cooper's integrative model (1999). In this version, the cognitive elements affected by dissonance may be personal standards (idiographic arousal) or normative standards (nomothetic arousal). In the case of soldiers, normative standards can turn into personal ones. On this basis, theoretically, they are highly sensitive to cognitive dissonance and thus to a potential cognitive attack. Indeed, cognitive dissonance incurred by psyops mobilizes soldiers' own standards to arouse dissonance-motivational discomfort-and make them change their behavior. These operations are highly effective when they decrease trust and certainty in soldiers' minds, thereby weakening the cornerstone of a properly functioning military, as we shall see in Part II.

With regards to the self-consistency model and Self Standard Model theory, the more a target is trained and conditioned, the more it can merge its professional values and self- perception, combining idiomatic and nomothetic standards. Consequently, training gives serious indicators on how soldiers think and react, thus helping in profiling. If the target is conditioned enough, we can postulate that this kind of profiling could be very effective and close to what a personal profile can be. Moreover, when reacting to a psyop built on cognitive dissonance, the target will not even be aware he had been influenced, as he changed his behavior according to what he perceived as the very values and elements which are part of his personality, as highlighted by our interviewees' responses to their perception of being a target. Consequently, in professional settings, compromised personnel will not be easily detected because they are still loyal to the overall pattern of what must be defended and what is worth fighting for.

It seems clear that the risk component of a deterioration in trust is a change in behavior based on a change in decision-making. But making a realistic prediction of what the agent behavior could be in order to secure an organization strategy (i.e., being sure that orders are going to be executed without modification) implies the supposition that decision-making is based on rationality. The "rational choice model" refers to theories of action in this area of research. Developed initially by Friedman (1953), the Theory of Rational Choice (TRC) is based on a holistic approach and has achieved paradigmatic status in economic sciences. More specifically, the TRC states that individuals act according to two criteria: maximization and coherence, in which rationality is a link between information and individual preference. While this theory helps predict expected decision results, some authors (Elster & Gerschenfeld 1986; Bourdon 2004; Allais 1955) highlight its lack of realism. TRC does not in fact take account of environmental factors, or of the subjects' experiences and intuition, and thus has certain inherent limitations.

In response to these, Kahneman and Tversky (1977) developed a decision model that departs from the theory of rational choice and attempts to define the mechanisms leading to individual and collective decision-making. Basing their theory on a heuristic approach that privileged heuristics and biases (HB), Kahneman and Tversky proposed decision-making as the product of mechanisms of interaction between the automatic system and the reflected system. In HB, mental process is constituted by two different systems: the automatic system (System I) based on immediately available knowledge, always able to generate an answer, and the reflexive system (System II) producing judgments and able to rationalize the ideas produced by System I.

More concretely, System I is based on innate abilities and makes quick associations between ideas (Morewedge & Kahneman 2010). It is known for translating emotion into impulsion, given that emotion is understood as a psychological and physical reaction to a situation. As such, System I may lead individuals to act suddenly (by reflex) without thinking about the consequence (Baratt 1993). System II uses memory and requires attention and effort to work (Beatty & Kahneman 1966; Kahneman, Tursky, Shapiro, & Crider 1969). Consequently, when this system is called into play, it becomes difficult for the individual to perform multiple actions at once (Kahneman 2012). In practice, System I generates suggestions (feelings or intuitions) when a new situation arises and submits them to System II, which decides to launch actions. But the two systems may conflict (Gilovich, Griffin, & Kahneman, 2002). In this case, automatic reaction coupled with the intention of correcting a situation may cause problems.

As to our subject of interest, cyber operations may attempt to change System I in order to produce unexpected behavior in the targeted organization's agents. Effectively, changing System I will generate biased suggestions and result in a change of actions triggered by System II based on System I data. Moreover, intuition-managed by System I- comes from emotions and thus from the affective domain. Here, military values may switch from cognitive to affective and be reinforced by individual experience, modifying the intuition. According to Klein (1999) intuition also comes from experience, and military personnel usually develop strong intuition in their own field of action. According to Kahneman's theory, the efficiency of cyber operation as described could be particularly important.

As such, this study examines the possibility that cognitive-dissonance-based psyops can change both soldiers' trust and behavior, as we will see in the presentation of data and analysis. Failure of frontline soldiers to carry out orders properly could endanger an entire operation. To cause this, cognitive dissonance must change a soldier's trust and decision- making. Accordingly, prior to proceeding we must look at the literature on trust and cohesion in military affairs.

Trust and Cohesion in Military Affairs

Trust seems to be the cornerstone of military cohesion and performance. According to McAllister (1995), trust has cognitive and affective underpinnings and is a factor of cohesion. Some meta-analytic studies show a positive relationship between cohesion and performance (Mullen & Cooper 1994). In military matters, performance means achieving the goal of a successful mission with as little loss as possible. Other sources propose three main components of military cohesion: relations between peers, relations between superiors and subordinates, and relations between the armed forces and the government (Stewart 1988; Etzioni 1961). Hence for soldiers, degrading the perception of government and increasing mistrust in superiors will deteriorate cohesion and performance, increasing the risk factor in mission success.

While trust is primordial, it remains a nebulous concept that needs clarifying. Bhattacharya et al. define it as "the expectancy of positive outcomes that one can receive based on the expected action of another party in an interaction characterized by uncertainty" (Bhattacharya et al. 1998). Another definition from Rousseau et al. (1998) indicates a psychological state comprising the intention to accept vulnerability based upon positive expectations of the intentions or behavior of another. Lastly, Mayer, Davis, and Schoorman (1995) define trust as "a willingness to be vulnerable to another party when that party cannot be controlled or monitored." These definitions are particularly apt here because our subjects live and act under extreme situations. Accepting vulnerability is very real in that they can be wounded or even die in action.

Previous definitions of trust described uncertainty as a game changer. For Rivolier (1998), extreme situations are fast-changing with high levels of uncertainty for participants. To be considered extreme, a situation must also be risky (Lièvre 2014). In military parlance, vulnerability is a recurrent keyword in the definition of trust. Given that soldiers operate in extremely risky situations, it seems clear that trust is a cornerstone in achieving their tasks. In the event of failure, over and above their own mission is the government's credibility and trustworthiness. This can be diminished in its citizens' eyes (if soldiers come home wounded or die in battle, for example) and possibly its allies' too. So, eroding the trust of soldiers may have considerable effects not only on the battlefield but also on the government's policy and stability.

Cyber Operations, Access to Information, and Changes in Trust

Finally, in terms of literature review, in the case of Ukraine, the target of cyber operations was information. Based on our observation of increased uncertainty resulting from cyber operations, a comprehensive update on research into this issue seems appropriate.

According to Negroponte (1995), we live in a digital world, and military personnel are no exception: they are as connected as anybody else. Many of the soldiers we surveyed are "Generation Y;"[2] judging by Internet and mobile data consumption figures, this generation is among the most highly connected in history. A survey conducted by Omnibus Institute in the United Kingdom showed that 18 to 34-year-olds look at their smartphone up to 100 times a day, in other words every 9 minutes and 50 seconds. The growth of smartphones has caused a major shift in how we inform ourselves. As far as social media consumption is concerned, mobile usage grew by 30 percent from 2015 to 2016, an increase of 581 million. Over 91 percent of the 2.8 billion people on social media are connected via mobile devices, and soldiers are no different.

---

[2] *Generation X* refers to people born between 1966–1976; *Generation Y*, or Echo Boomers or Millennials, to people born between 1977–1994; *Generation Z* to people born between 1995–2012.

Cyberspace, being a vector of information, changes how people inform themselves, and therefore influences their trust. Although information access may vary depending on age, the trend is clear: more and more people use the Internet to get information, many through social networks. Only two soldiers in our sample were not connected to social networks. So, people can easily be targeted in cyberspace by "information" created for the need of a psyops strategy using cognitive processes. Given that all the soldiers interviewed owned smartphones and said they use them for surfing the web and social networks, analyzing the trust attributed to web-sourced information seems important in understanding how it can be used by an adversary to influence soldiers' perceptions. According to the American Press Institute and the Associated Press-NORC Center for Public Affairs Research (2016), the person sharing a post was considered more important than the original source of information. The survey results showed that the individual sharing the information had a major effect on how trustworthy the information was considered: 51 percent of people said an article was well-reported when shared by a person they trust. With Facebook, 48 percent of the interviewees said how much they trusted the person posting the article influenced how much they trusted the information in question.

As to the military, given the special training and resulting team spirit, we postulated that if information is posted by a soldier, or someone sharing the same values (volunteer, etc.), a soldier's trust in the content will increase. The analysis of data in Part II vindicate this.

The Case of Ukraine

Ukraine has been facing the biggest deployment of cyber operations against a country ever seen. In light of the literature presented in Part I, we examined the possibility that the trust of active and veteran Ukrainian soldiers had been affected by cognitive methods of Internet-based attacks. The purpose was to see whether their perceptions had altered and uncertainty increased in relation to their Internet consumption, and if so whether the effect was significant enough to modify their behavior. Another key question was whether they were aware of being potential targets or not.

Significance of the Case Study

Before answering these questions by means of data analysis, we discuss the value of Ukraine as an example. In Ukraine, the kinetic conflict was accompanied by massive waves of cyberattacks of many kinds. While Russia had already used such methods against Estonia and Georgia, the case of Ukraine is unprecedented in its magnitude. For many experts (Greenberger 2017; Weedon, 2015), it was a blueprint for testing various methods and estimating their effects. For NATO's CCDCOE, the use of cyberattacks by (in all appearance, Russia) was part of a broader strategy of information warfare (Geers et al. 2015).

Specific to Ukraine, cyberattacks included psyops cyber operations intended to affect as many people as possible and weaken the central government and country. We found that various channels were used in order to match the psychological profiles of targets and that the military was no exception, quite the contrary. Many military personnel received text messages on their mobile phones, encouraging them to stop fighting, go back home, or leave their positions. Meanwhile, more subtle operations were conducted through the Internet. We found that the soldiers we interviewed had received fraudulent messages. For example, the profiles of dead soldiers were used to send Facebook chat messages and spread "information" to sow doubt and mistrust toward state institutions. The soldiers explained that since they didn't know their "brothers in arms" were actually dead they were more inclined to believe these messages. This is paramount as such operations used soldiers' in peer-group trust to sow doubt about members of government or commanding officers.

Active fighters were not the only targets: a vast operation was also conducted on veterans. Veterans are ideal targets, as they exemplify an ideal of heroism for the general population, occupy a special position as thought leaders or national symbols, and have been trained in firearms handling. Many of them are young and thus part of the labor force and have political weight as voters and/or potential political leaders. Moreover, their past in EO makes them a potentially explosive social group. The way they feel can therefore have a profound influence on people's perceptions.

Except for certain occasions-tactical war phases, for example-the final goal seems to be to increase uncertainty in the minds of soldiers and fighters. To this end, some information media such as blogs or social networks were used (Weedon 2015). These practices are in accordance with Russian cyber warfare strategy such as developed by General Gareev (2015), or Colonel Chekinov and Lieutenant General Bogdanov (2011, 2013). According to them, information is a component of New-Generation War (NGW). As such, subversive operations involving information can be used to create chaos and provoke various kinds of disturbance, including weakened state resilience. Russian cyber warfare strategists have even recommended the use of mass media to stir up chaos and confusion in government but also in military management, particularly command and control. This matches the desired results of operations conducted in Ukraine. As a mirror effect, their recommendations for safeguarding Russia from such operations can be reversed for devising battle plans. For instance, their proposals included keeping sources of domestic (Russian) information out of reach of adversarial influence; yet in Ukraine, information channels were penetrated by exogenous influence.

Last but not least, Russian cyber warfare strategists have advised that information, including psychological warfare, should predominate in NGW, and be used extensively. The effect on the human psyche targeted is clearly to misinform, as well as encourage discontent and unlawful acts. Analysis of the data collected in Ukraine thus seems to suggest a global military strategy, since loss of trust resulting in disengagement and changes to soldiers' behavior can be assimilated with unlawful acts based on discontent. Further, the result of these operations upon the execution of orders is clearly disruptive of military management, and command and control.

Data Collection and Research Methodology

Since 2014, we have made several trips to Ukraine, during which we witnessed cyber operations on various population levels. The data collection presented is based on research into the literature and field observations, static and dynamic, on and off the battlefield.

The goal of the research trip in February 2017, during the battle of Avdiivka, was to interview soldiers and veterans and establish whether their uncertainty had changed according to their Internet consumption, specifically according to a certain type of content. For context, the government had been vigorously promoting military values since the beginning of the conflict in Ukraine. Travel was initially planned to Avdiivka. However, due to the battle that broke out in the city, safe access was not assured, and we traveled instead to Poltava and Krementchuk. Poltava has a hospital where many wounded soldiers were treated for physical and psychological trauma. Krementchuk has a military school and a large part of its population was affected by the war. In addition, many soldiers were based in these cities to ensure turnover on the front line, ensuring ease of access to interview subjects.

Survey methodology

After local contacts had introduced the researchers to various groups, thirty active soldiers and veterans were interviewed after being divided into two groups, one of fifteen veterans (all wounded physically or psychologically) and one of fifteen active soldiers. The soldiers were assured of their anonymity and that the interviews would be neither transmitted nor published, as much for reasons of national security as to increase their trust in the researchers. The surveys used a qualitative approach designed to obtain as much information as possible.

Concerning the question set, the interviews were not strictly directed but many questions were recurrent, thus allowing an analytical framework to be defined. For instance, the soldiers were asked to quantify their trust in various structures (group, institution, and government) on a Likert scale from 0 to 5, as well as their confidence in various kinds of media (radio, television, social networks, and alternative media such as blogs and nongovernmental information websites). They were also invited to explain their Internet habits and how they get their information. They were surveyed on how they felt about their institution and government before and after operations and also asked about the various media they listen to and why. The subjects were thus interviewed to ascertain whether their trust and perception of the government and their military hierarchy could have been influenced by cyber content and/or their Internet consumption. Lastly, they were asked about their perception of being or not being a potential target for psyops on the Internet. They were asked to elaborate on their answers. The interviews were sometimes followed by discussions on Skype or by email at the soldiers' discretion. Several of them used these discussions to send material, especially fake profiles they had spotted, or blog posts that particularly interested, impressed, or influenced them.

Analysis and Results

The qualitative data were processed with Nvivo 12; some data were extracted using Excel in order to express results as percentages, allowing us to perform the analysis presented here.

Analysis of the data collected from interviews with soldiers in Ukraine sheds light on several points. First, their daily consumption of Internet data was very diverse (from 0.5 to 6 hours). Nonetheless, it seems that the ones who changed their mind the most were those connected two to four hours per day, in other words it was not the biggest consumers who were the most influenced. However, we need to look more deeply into this observation, especially when it comes to the data consumption environment (private or public, etc.). Second, in terms of age, it seems that soldiers belonging to "Generation Y" were more responsive and increased their uncertainty level toward institutions and government most. Some of them went so far as to say they had changed their behavior enough to not strictly obey a direct order. As justification, they explained that they had discovered details revealing that some of their commanders were not as "trustworthy" as they expected them to be. One of them explained that "I won't risk my life for someone who shows no respect for us, the sons of Ukraine, and for our wounded." Here, we also noticed the development of a gap between "us" the soldiers, and "they," "the others," in other words, the military hierarchy or members of government.

The interviewed soldiers' experience and participation in kinetic operations seemed to give them the feeling that they were more legitimate than the government in judging the merit of an action. For instance, one explained, "I believe less and less in people who do not have their hands in grease, who had never been on the field;" another told us that "there are too many people talking about what they do not know ... it's getting on my nerves." Yet another soldier concluded that "they are wearing figures just for the pictures, reality is that they don't know anything." At the same time, one thing we observed in a focus group of Ukrainian soldiers was that they never stopped believing in the core of their cause ("fighting for Ukraine") but have gradually dissociated it from Ukraine's government. One explained, "I believe less and less in our government and its capacity to help us. I only believe in Ukraine, those who fight. Those who do not stop fighting."

Indeed, the growing mistrust of the "others" has ratcheted up doubt in their government and, for some, President Poroshenko in particular. As to their level of trust in various media, they mainly seemed to trust social networks and alternative media more than official ones, which are perceived as tied too much to government or oligarchs. For instance, one explained to us that "the newspapers, the big media belong to men who serve politics, so it's not the truth that they want us to believe." Another said, "when you are not paid for your publication you are more honest, so I think blogs and social network lie less than big media."

In explaining this gap in trust, soldiers argued that they mostly trusted their brothers in arms" and volunteers because they had similar experiences and fought for the same cause. It also seemed quite clear that their self-perception was strongly linked to their soldier's identity (even if they had other professions before the war) and patriotic fight for their country. Furthermore, they described themselves using military values and vocabulary (like patriotism, honor, and loyalty). It seems that they adopted these cognitive data (learned by training, the very nature of their job) as part of themselves, of the way they perceive themselves: these cognitive elements have become personal and affective ones. The cognitive components of their training and conditioning eventually became part of their personality.

According to the research we conducted, the soldiers we interviewed and surveyed seem to have been targeted by multi-step operations. Firstly, strong cognitions inherited from training were reinforced. In a military environment, this was easy to achieve since the targets believed their contact (the writer of the papers they read or sharer of the information seen or heard, etc.) was involved in military operations and therefore shared the same values. This was indicated by the use of expressions such as "brother in arms," "sons of Ukraine," "we," or "us."

Furthermore, since this concerns values of importance to soldiers, the agent involved appears increasingly trustworthy. Consequently, an adversary can send misinformation to his target, linking this cognition with the field in which he wants to sow doubt.

For instance, soldiers showed us articles associating a member of government or a superior with the disrespectful treatment of veterans or wounded fighters who, in a soldier's belief system, deserve military honors. As the information is completely opposed to the way soldiers want to see themselves and their values, this content was rejected. This example is no exception, and this kind of reaction is founded on soldiers' self-perception, even if it is based on cognitions resulting from training and conditioning. We thus noticed that some soldiers' strong cognitions were reinforced, particularly via Internet chats with apparently fake profiles or through the fraudulent use of deceased soldiers' profiles. Other strong cognitions were increased through blogs forwarded by "friends" or by other soldiers, sometimes real but sometimes created on purpose.

Thanks to mobile data, interactions between the targets and psyops agents can occur every day. Some may encourage targeted persons to perform specific actions at specific times, but this is very unlikely. It would require the conjunction of two complicated phenomena: an exact action and an exact time. When the focus group was asked whether they thought they could be a target of interest, the answer was unanimously "no." The explanation given was that, in their opinion, they were not strategic targets because they were not officers and did not hold strategic information or data. They particularly did not feel they could be strategic targets themselves. On the other hand, the focus group agreed they could be targeted by propaganda, but as Ukrainians and not as soldiers. Were it to happen, however, they thought it wouldn't work on them as they "knew separatists' and Russian lies too well to be fooled." Yet, the focus group only referred to "higher" propaganda and not its most subtle forms, i.e., the use of cognitive dissonance. One counterintuitive result is that soldiers felt they were not affected by cyber operations, all the while acknowledging that their opinion about the government and military command had changed as they "learned" more about them, largely through information obtained online.

The implications here are important. If the perpetrator of an attack were to obtain an increase in uncertainty among a large enough proportion of a military team, this could increase his ability to cause the operation to fail. With increased levels of doubt and uncertainty, military personnel may not behave as they are supposed to. This is very dangerous as mission planning is built on the basis that soldiers will obey orders without variation and relies on behavioral stability. In the end, the nature of behavioral change does not matter. What does is that overreaction could be as dangerous for the conduct of a mission as underreaction. By underreaction, we understand a weaker response than expected, such as not executing an order or taking more time to carry it. In all cases, targeted soldiers do not accord with their conditioning and engagement. It is this 'disengagement' that can jeopardize the mission. These operations and their observed effects are related to trust in military teams, and to military psychology. Using cyber operations could be particularly advantageous for an adversary. Their efficiency can be significant on two levels, first from a practical point of view, in other words, on efficiency, and second in terms of the interest to the organization using these kinds of operations, in other words, on strategy. Concerning efficiency, many people use the internet to get information, in other words, cognitive elements. Cognitive dissonance can thus be used to trigger a psychological motivational discomfort and attempt to change the behavior of any subject. In addition, the internet can be used all day long to affect people, which means they can be subjected to several phases of cognitive dissonances, thereby increasing the rate of success in behavior change. Moreover, in accordance with the theories of Aronson (1968, 1992, and 1999) and Cooper & Stone (1999), cognitive dissonance can be deployed if a target's training is known and the values learned have become part of their personality, as occurs in the military. The Ukrainian case seems to validate this theory.

Conclusion

Based on the above observations, it appears that cognitive dissonance through misinformation circulated on the internet can be used as an efficient tool for psyops to affect trust. If Ukraine were indeed a blueprint for Russian destabilization operations, the methods used were presumably worked out beforehand. In agreement with the concept of cognitive dissonance, the more the cognitive and affective dimensions are merged, the easier it should be to influence somebody using methods discussed in this theory. In the case of soldiers, since training and personal values merge in accordance with self-consistency theory, an adversary can easily obtain the profile of a military team, including soldiers' self-perceptions. In other words, the more someone is trained and conditioned, the more he can become a target. This holds the possibility of many practical consequences.

First, using the training and conditioning of military personnel as key component of these cyber operations, the assailant does not need to send agents into the field, close to targets, to perform individual psychological profiles. It is not only safer, decreasing the risk of being unmasked, but also less expansive, as long as the training and values inculcated have been studied enough. Looked at cynically, using this method, part of the work is done by the victim, not by the assailant, thereby reducing the latter's work. Further, as the target changes his behavior in accordance with the values he believes in and the training he received, he will not realize he is acting under someone else's influence. In our study, soldiers stated they couldn't be targeted and, even if they were, wouldn't be affected. Finally, we may suspect that when a soldier is subject to long-term cognitive dissonance, even if he does not change his behavior, he will feel psychological discomfort, for cognitive dissonance affects brain activity, in particular the prefrontal cortex and anterior cingular cortex (Gehring et al, 1993; Amodio, et al, 2004; Harmon-Jones et al, 2008). This in turn becomes a human management issue during both the short and long terms for the assaulted person. On the strategic level, the benefit of this kind of weaponization means that it will be very complicated to prove the involvement of the head of operations, and just as complicated to detect the attack itself. In the Ukrainian case, even if president Poroshenko points the finger at Russia for many of the cyber-attacks and cyber operations his country has faced since 2014, actually proving Russian involvement is very complicated. Given the wide range of people and events involved, it would be very difficult to link attacks conducted via multiple channels with an overall state strategy. Concerning Ukraine, various hacker groups are known to be involved in the war, such as CyberBerkuts, Cozy Bears,

FancyBears, or SandWorm. However, despite strong suspicion, no sufficient proof has ever been brought to a court of law. Consequently, any government might find it useful to hide behind seemingly autonomous hacker groups to target another country without fear of falling under the scope of international law, military or otherwise. Alternatively, those attacked can hardly respond in kind if they wish to respect the law.

Further, since either detecting or attributing these cyber operations is complicated, assailants could more than likely apply them during peacetime. The high level of internet penetration and the fact that cyber's multi-channel nature can be used to generate cognitive dissonance and affect System I enough to upset its relation with System II is extremely interesting, particularly because no commitment is needed for the outbreak of information dissonance (Vaidis and Gosling; 2011). Thus, if information that appeared on a smartphone can trigger dissonance, the target does not need to have sought the information for a reaction to be emulated. Moreover, the rapid growth of 5G will probably play its role in future developments, revolutionizing our relationship with technology and information. As such, the present results are probably just the first step in further research, as operations like this are likely to increase in importance as time progresses.

Bibliography

Adams, B., Webb, R., Trust development in small teams, Defense research and development Canada, Toronto contract report, CR-2003-016, Guelph, Ontario; Human System Incorporated, 2003.

Allais, M., Fondements d'une theorie positive des choix comportant un risque et critique des postulats et axioms de l'école americaine. Paris: Imprimerie Nationale, 1955.

Amodio, D., Harmon-Jones, E., Devine, P.J., Curtin J.J., Hatley S., & Covert A., 'Neural signal for the detection of unintentional race bias', Psychological Science, 2004, pp. 85-93.

American Press Institute and the Associated Press-NORC Center for Public Affairs Research, 'Who shared it?' : How Americans decide what news to trust on social media, American Press Institute, 2017.

Aronson, E., 'The cognitive and behavioral consequences of the confirmation and disconfirmation of expectations', Application for research grant submitted to the National Science Foundation, 1968, Harvard University.

Aronson, E., The return of the repressed: dissonance theory makes a comeback, Psychological inquiry, 1999, pp. 303-311.

Barratt, E.S., Impulsivity: Integrating cognitive, behavioral, biological, and environmental data in W. G. McCown, J. L. Johnson, & M. B. Shure (Eds.), The impulsive client: Theory, research, and treatment, 1993, pp. 39-56. Washington, DC, US: American Psychological Association.

Battacharya, R., Devinney, T.M., Pillutla, M.M., A formal model of trust based on outcomes, Academy of management review, 1998, 23(3), pp. 459-472.

Bourdon, R., Theorie du choix rationel ou individualisme méthodologique? Revue du MAUSS, 2004, 2, pp. 281-309.

Cabinet Deloitte, Usages Mobiles 2015: A Game of Phones, annual report, January 2016.

Chekinov, S.G, Bogdanov, S.A, "The Nature and Content of a New-Generation War," Voennaya Mysl' (Military Thought), 2013, No. 10, pp. 13-25.

Clarke, A.R., Cyber War, Harper Collins, 2010.

U.S DoD, JP1-02: Department of Defense Dictionary of Military and Associated Terms, DoD, Washington: DoD., 2010. https://fas.org/irp/doddir/dod/jpl_o2.pdf.

Elster, J., Gerschenfeld, A. Le laboureur et ses enfants: deux essais sur les limites de la rationalité. Paris: Les Editions de Minuit, 1986.

Etzoni, A., A Comparative Analysis of Complex Organizations on Power, Involvement, and their Correlates, The free press, New York, 1961.

Eurostat statistics explained. internet activities in the past three months by age group EU-28, 2016, Eurostat edition, 2016. http://ec.europa.eu/eurostat/statisticsexplained/index.php.

Festinger, L., Une théorie de dissonance cognitive, Les classiques de sciences usines et sociales, Paris, Enrick B Editions, 1956.

Friedman M., The methodology of positive economics, *in* Friedman, M., *Essays on Positive Economics*, Chicago, University of Chicago Press, 1953.

Gareev, M.A (unattributed report summarizing his speech), "How Does One Develop a Modern Army?" Krasnaya Zvezda Online, 2016, 11.

Geers, K., Giles, K., Wirtz, J.J., Lewis, J.A, Lebicki, M., Koval, N., Pakharenko, G., Weedo, J., et al, Cyber War in Perspective: Russian Aggression against Ukraine, CCDCOE, 2015.

Gehring, W.J., Goss, B., Coles, M.G.H., Meyer, D.E., & Donchin, E., A neural system for error detection and compensation, Psychological Science, 1993, pp. 385-390.

Gilovich, T., Griffin, D., & Kahneman, D., Heuristic and biases: the psychology of intuitive judgement, New York, NY, US: Cambridge University Press. 2002.

Greenberg, A., Hunting the hackers: How Ukraine became Russia's test lab for cyberwar, Wired, (20 June 2017).

Harmon-Jones, E., Brehm, J., Greenberger, J., Simon, L. & Nelson, D., 'Evidence that the production of aversive consequence is not necessary to create cognitive dissonance', Journal of Personality and Social Psychology, 1996, pp. 5-16.

Harmon-Jones, E., Harmon-Jones, C., Fearn, M., Singleton, J.D., & Johnson, P. 'Left prefrontal cortical activation and spreading of alternatives: tests of the action-based model dissonance', Journal of personality and social science, 2008, pp. 1-15.

Hootsuite & We are Social, Special Reports Digital in 2017: Global overview, January 2017. https://wearesocial.com/special-reports/digital-in-2017-global-overview.

Jonas, E., Greenberg, J., Frey, D. Connecting terror management and dissonance theory: evidence that mortality salience increases the preference for supporting information after decision', Personality and social psychology Bulletin, 2003, pp. 1181-1189.

Kahneman, D., & Tversky, A. (1977). Intuitive prediction: Biases and corrective procedures. DTIC Document. Retrieved December 3, 2016.

Kahneman, D., & Tversky, A. (1981). The simulation heuristic. DTIC Document. Retrieved December 8, 2016.

Kahneman, D., Beatty, J., Pupil diameter and load memory, Science, 1966, vol 154, pp.1583-1585.

Kahneman, D., Tursky, B., Shapiro, D., Crider, A., Pupillary, heart rate, and skin resistance changes during a mental task, Journal of Experimental Psychology, 1969, 79, pp.164-167.

Klein, G.A., Source of Power: How People Make Decisions. Cambridge, Mass: MIT press, 1999.

Kulpin, A., Rauscher, K.F., Yaschenko, V., for East-West Institute, 2014, Critical Terminology Foundations 2, In: Russia-US Bilateral on Cybersecurity, Habes B. Godwin III (eds), Policy Report 2/2014.

Libicki, M., Cyberdeterrence and Cyberwar, RAND Corporation, 2009.

Lièvre, P., Repères pour un management des situations extrêmes, 25e congrès de l'Association francophone de Gestion des ressources humaines, 6 & 7 November 2014.

McAllister, D.J., Affect- and cognition-based trust as foundation for interpersonal cooperation in organizations, Academy of Management Journal, 1995, pp. 24-95.

Mayer, R.C., Davis, J.H., and Schoorman, F.D., 'An Integrative Model of Organizational Trust', the Academy of Management Review, 1995, Vol. 20, No. 3, pp. 709-734.

Morewedge, C.K., Kahneman, D., Associative processes in intuitive judgement, Trends in cognitive science, 2010, vol 14, Issue 10, pp. 435-440.

Mullen, B., Cooper, C., 'The relation between group cohesiveness and performance: An integration', Psychological Bulletin, 1994, pp. 210-227.

Negroponte, N., Being Digital, Alfred A Knops, New York 1995.

Orléan, A., Hétérodoxie et incertitude, Epistémologie et Autonomie, Les cahiers du CREA Ecole Polytechnique, 1986, Paris.

Rivolier, J., Stress et situation extrêmes, Bulletins de la psychologie, 1998, pp. 6-20.

Rousseau, D., Sikins, S., Burt, R., and Camerer, C.F., 'Not so different after all: a cross discipline view of trust', Academy of management review, 1998, 23(3), pp. 393-404.

Schaap, A.J., 'Cyber warfare operations: development and use under international law', Air Force Law Review n°64, 2009, pp. 121-170.

Stewart, N.K., The south Atlantic conflict of 1982. A case study on military cohesion, U.S Army Alexandria, 1998.

Stone, J., Cooper, J., A self-standards model of cognitive dissonance, Journal of Experimental Social Psychology, 1999, pp. 228-243.

Vaidis, D., Gosling, P., 'Is commitment necessary for the arousal of information dissonance?', Revue Internationale de Psychologie Sociale, 2011, n°23, pp. 33-63.

Weedon, J., Beyond 'Cyber War': Russia's Use of Strategic Cyber Espionage and Information Operations in Ukraine, Chapter 8 In: Kenneth Geers (ed.), Cyber War in Perspective: Russian Aggression against Ukraine, NATO CCDCOE Publications, Tallinn, 2015.

How to Prevent Motorbike Attacks in the Sahel?[3]

Thomas Flichy de La Neuville
(École spéciale militaire de Saint-Cyr)

Abstract

Widely spread in the Sahelian zone for reasons having to do both with economy and mobility, motorcycles have become an element of social prestige for youth in the region. At the same time, their use has been partially diverted for criminal purposes, leading to an increase in swarm attacks. This may be the most recent manifestation of an ancient nomad technique employed against sedentary populations. Facing this danger, should we attempt to hinder these motorcycle by "strangling" their circulation or, on the contrary, counter them with an even greater mobility? This essay discusses the origins of the problem, describes the ways in which it manifests itself, and considers possible responses.

Motorcycles in Africa, a Versatile and Popular Tool

Fluid, concealable, popular and cheap, motorcycles have spread throughout Africa over the last thirty years. Their uses are multiple: 1.5 million motorcycles are used on the continent as taxis, transporting passengers and goods, particularly in rough terrains. Furthermore, farmers have been using them to deal with the problem of having to travel back and forth as they irrigate their fields; a simple pump run by a motorcycle has proven an innovative solution in this case.

[3] I would like to thank the following cadets of the French military academy for their assistance in conducting the research for this paper: Cadets Bendaravicius, Cairoli, Calamiello, Castaldi, Castier, Colas, Daillet, Dhalluin, Girard, Jonuska, Kastner, Koga, Maillard, Saccoman, and Selleri.

Since the 1990s, a whole culture based on motorcycles has developed in the cities of northern Nigeria and Cameroon. In a market that was at first dominated by the Japanese motorcycles, the popularity of particular models varies by city, while locals apply nicknames reflecting their own cultural preferences. For example, in Cameroon, the Honda CG125 is called "Bazooka," the Suzuki A100 "Chagari" and the Suzuki AS100 the "black cat." These Japanese motorcycles were very popular until the arrival of their Chinese competitors in 2004, including Nanfang, Lifan, Jingseng, Carter, and (the fastest) TVS125. Equipped with a lengthened seat, these latter can carry three to four riders at a time, as well as up to six plastic cans of sixty liters each, convenient for gasoline smuggling. Their gas mileage is low, but this disadvantage is offset by the purchase price, which is 50 percent less than that of Japanese vehicles, making it possible for young people to acquire a motorcycle by saving up, without relying on help from their families.

Giorgio Blundo has studied the circuits of distribution of the Chinese motorcycle in Africa.[4] According to his findings, the city of Cinkassé, in the north of Togo, appears to be a dry port for the import of Chinese motorcycles and the biggest market for the vehicles in western Africa. Situated on the border with Burkina Faso, Cinkassé has become a center to which buyers come from the whole of the Sahel. In Togo, Chinese motorcycles are delivered in spare parts before being assembled by African workers benefitting from the assistance of Chinese technicians.[5] Because of the very modest cost of the Chinese motorcycles, local production is almost impossible.[6]

Before moving on to the security problems posed by motorcycles in Africa, their importance in the context of that continent may be seen in

---

[4] Giorgio Blundo is an anthropologist and is presently Directeur d'Etude at the Ecole des Hautes Etudes en Sciences Sociales (EHESS). Since February 2016, he has been in Togo, where he has conducted research on the project "Chinese Motorcycles in Africa, Globlalization, Mobility, and Politics (Togo, Benin, and Ghana)."

[5] Yishuang Liu et Wai Shan, "L'importation des motos chinoises a permis une circulation technologique au Sahel," French.people.cn, October 7, 2017.

[6] Paul Starkey, *The Rapid Assessment of Rural Transport Service*, report prepared for Intermediate Technology Consultants, March 2006, 2.

a few illustrative examples. For instance, the vehicle serves all of the commercial circuits of smugglers (known as "stuntmen") in the Mandara mountainous region of Cameroon. It provides easy access to suburbs via serpentine mews, and has affected even small villages, opening them up to the wider world and allowing for easier seasonal movement of livestock. One could in this sense see the motorcycle in Africa as a tool of spontaneous development; unfortunately, it has also become an engine at the service of armed criminals.[7]

The Operating Mode of Motorcycle Combat: A Trio of Fighters

Motorcycle banditry is a widespread phenomenon that can be found from the Balkans[8] to Afghanistan.[9] It can take the form of assaults on police[10] or gendarmerie[11] stations by a few armed men on motorbikes; sometimes motorized bands surround villages in order to loot them,[12] resulting in victims among both civilians and representatives of law and order.[13] Motorcycle attacks may also involve several men armed with AK-47s committing a kidnapping,[14] storming a marketplace,[15] or successfully carrying out a suicide attack.[16] In some cases, hundreds of fighters erupt

---

[7] Christian Seignobos, "Boko Haram: Innovations guerrières depuis les monts Mandara," *Afrique contemporaine*, no. 252 (April 2014): 149–69.

[8] See the work of professor Alfredo Nunzi, from 2018.

[9] Colonel Jean-Louis Dufour, "Terrorisme en essaim," *Revue L'économiste*, no. 2981, March 13, 2009.

[10] "Aguel-hoc: Les casques bleus visés par une attaque terroriste," *Studio Tamani*, April 6, 2018.

[11] Human Rights Watch, "Abus commis par des islamistes armés et par des membres des forces de sécurité au Burkina Faso," May 21, 2018.

[12] Seignobos, "Boko Haram: Innovations guerrières."

[13] Bamanda.net, "*GAO: Trois individus tués par un groupe armé sur six motos*," June 29, 2018.

[14] Human Rights Watch, "Abus commis par des islamistes armés."

[15] Seignobos, "Boko Haram: Innovations guerrières."

[16] Robert Johnson, "*La campagne de contre-insurrection en Afghanistan*," *Revue Stratégique* no. 103 (2013).

seemingly out of nowhere on motorcycles, commit an assault lasting only a few minutes, and suddenly disappear.[17]

The most frequent operating mode is as follows. Seated on the fuel tank, the driver holds the slightly raised handlebar while a second passenger seated on the metal extension of the seat at the rear takes the role of gunner. A third passenger, in the middle, points out targets and provides the gunner with ammunition. Modifications to technique are employed depending on local conditions. For example, Mandara "stuntmen" have adapted their Chinese motorcycles to the mountainous terrain, giving them double rear shock absorbers and protecting their inner tubes from the needles of Balanites trees with a metallic ribbon.[18]

Motorcycles are also used as tool in the recruitment of young men into jihadist movements. For example, in Burkina Faso, participation in jihadism enables simple shepherds to get access to weapons and to a motorcycle.[19] But it is the case of Boko Haram that is particularly instructive.

Two elements are favored by Boko Haram in attracting young men into its ranks: the gift of a motorcycle and the offer of one or several women. These two elements are far from trivial. In fact, obtaining a motorcycle is both a condition of spatial mobility in the regions affected and a form of social recognition. For instance, after the proclamation of a state of emergency and the prohibition of motorcycles in Diffa, the promise of a motorcycle has become an even greater factor of attraction for the local youth. In the minds of these young recruits, Boko Haram gives them that which the Nigerian state and society no longer offer.[20]

Indeed, the Chinese motorcycle has become inseparable from Boko Haram. It is present in all of the group's actions: attacks, kidnappings, rapid strikes, raids, and open battle with the army. After the fighting, photographs published in the press always show the ground

---

[17] "A Swarm of Motorbikes, Then Heavy Fire: Witnesses Shed New Light on Niger Attack, " *The Guardian*, November 15, 2017.

[18] Seignobos, "Boko Haram: Innovations guerrières."

[19] Associated Press, "In Heart of West Africa, Burkina Faso Faces Rising Extremism," August 15, 2018.

[20] Mathieu Pellerin, "Les trajectoires de radicalisation religieuse au Sahel,"*Notes de l'Ifri*, (February 2017): 26.

strewn with burnt motorcycle carcasses.[21] The entire logistics of Boko Haram are based on packs of motorbikes. In each motorized unit, one usually find a mechanic, a repairman, and an artificer responsible for ammunition and explosives; the bike itself belongs either to the fighter or the squad leader. In practice, the two-wheeled vehicles are used for small, rapid strikes but also enable the quick concentration of hundreds of fighters, the encirclement of a village, and rapid dispersion. Furthermore, attacks combining Chinese motorcycles, Kalashnikovs, and the mobile phone have proved particularly efficient.[22] The following and other actions of Boko Haram in Nigeria, as in Cameroon, were perpetrated thanks to this combination: Waza on May 17, 2014, Izage soon after, Kolofata and Hile-Alifa on July 26, 2014, the capture of the big market of Banki in August 2014, Ngala in September 2014, among others. The same may be said for the group's hostage-taking activities.[23]

Making Good Use of an Ancient Swarming Tactic

For millennia, swarming attacks have been conducted by nomads against sedentary farmers. For example, this was a favorite tactic of the Scythians. During the siege of Samarkand, Spitamenes used swarming against the army of Alexander the Great. Small groups of horsemen would appear, empty their quivers, then disappear to resupply as another group came in. Reacting to this effective tactic, Alexander had to split his army and attack the enemy supply line, forcing them into hand to hand combat.[24] Later, the Parthians crushed the Romans at the battle of Carrhae, using mobile tactics to saturate the enemy and prevent them from being able to maneuver.[25] The Mongol invasions provide another example of swarming attacks, but on a far larger scale. Indeed, the Mongolians relied on the combination of mobility and intelligence to rapidly gather together several armies by making them pass by areas where the enemy was not present, in order to then attack en masse at a previously agreed upon location. A large network of messengers capable of traveling quickly over great distances

---

[21] Seignobos, "Boko Haram: Innovations guerrières."
[22] Seignobos, "Boko Haram: Innovations guerrières."
[23] Seignobos, "Boko Haram: Innovations guerrières."
[24] Sean J. A. Edwards, *Swarming and the Future of Warfare*, RAND, 2004, 183.
[25] Edwards, 185.

to collected intelligence enabled the Mongols to keep the initiative on an almost permanent basis.

The development of urban warfare in the twentieth century necessitated the evolution of military swarming tactics. Thus, in the battles of Grozny between Russians and Chechens (1994, 1996, 1999), the insurgents defended the city by quickly infiltrating the enemy rear via tunnels, ruins, and constructed passages in order to brutally engage the enemy by surprise and at a short distance, before breaking contact and retreating. The Russians lost many tanks destroyed by RPG, saw their positions threatened, and were defeated during the first and second battles of Grozny. They ended up setting up a "spider's web" made up of advance outposts in order to limit the enemy's mobility. Despite these efforts and overwhelming Russian numerical superiority, the Chechens inflicted heavy losses on the Russian army, frequently encircling the outposts.

Swarming has also been used in terrorist attacks. For example, during the bombings in Mumbai (2008, 179 dead), five teams of two men each led simultaneous attacks against several objectives. It should be noted that the United States has been strategically preparing for enemy naval swarm attacks for two decades now. Thus, *Exercise Millenium Challenge 2002* opposed an Iranian-style "red force" commanded by the American general van Riper and using a flotilla of small boats against a conventional "blue" naval force. This simulation resulted in the destruction of sixteen warships, including an aircraft carrier, two helicopter carriers, and five troop transport vessels with 20,000 men on board. Today, Western armies are working on the formation of swarms of drones.[26]

"Strangling" the Enemy or Fielding an Even More Mobile Response?

In the face of a swarm of enemies, responses have been varied. The most common counter tactic consists of quashing the swarm by strangling its supply chain. Alexander did just that during his siege of Samarkand. Some African states have attempted to use similar tactics to hinder biker gangs by forbidding the circulation of motorcycles at night.

[26] Scott Simonsen, "How Swarm Intelligence Is Making Simple Tech Much Smarter," *Singularity Hub* (2018): 1–5.

As far back as 2011, the local government in the Nigerian federal capital city of Abuja instituted a ban on *mototaximen Okada*, out of concern that they could be used by Boko Haram to commit targeted murders.[27] In Cameroon, since 2013 the threat from Nigerian motorbikes was so great that the government imposed a curfew from 6 pm to 6 am on the northern border, from Mandara to Am Chidé and Banki. Since then, this measure has been applied to all the regions of Cameroon north of Maroua.[28] In Togo in 2014, the national army was ordered to seize all unregistered motorbikes.[29] In September 2018, Burkina Faso, which is facing recurrent jihadist attacks in the east of its territory, banned two- or three-wheel night traffic in this regions; since then, the circulation of motorbikes is officially not allowed from 7 pm to 5 am.

The second response consists in copying the enemy and in outclassing it with higher levels of mobility. For this reason, the Iranian counter-riot policemen are equipped with motorbikes.[30] For their part, Lithuanian Special Forces in Afghanistan have used powerful Yamaha motorbikes, going so far as to build a special training camp in Lithuania on uneven grounds. Use of the bikes in Afghanistan allowed the Lithuanians to avoid traps set for heavy tanks. There have been some limitations to this tactic, however. For instance, in motorbike fights the Afghans were certainly slower than the Lithuanians, but also lighter because their bikes were not weighed down with body armor and weapons.[31] Furthermore, in light of the use of motorbikes by special forces, ambushes in Afghanistan began to be displaced to 1.5 km away from the main roads.[32] Finally, in recognition that motorbikes are not by nature discrete vehicles, being loud, the Defense Advanced Research Projects Agency is currently developing a new two-wheeled vehicle called

---

27 Seignobos, "Boko Haram: Innovations guerrières."
28 Seignobos, "Boko Haram: Innovations guerrières."
29 K. Mensah, *L'Alternative-Togo*, May 24, 2014.
30 Mohammad Javad Shafiei, "Affronter les forces de l'ordre en Iran l'année 2009," *Revue Outre-Terre* no. 28 (2011): 273–76.
31 Seth Robson, "Soldiers on Motorcycles Might be Most Effective in Afghanistan," *Stars and Stripes*, June 8, 2013.
32 "Why Aitvaras Motorcycle Soldiers Cause Fear for Taliban,"*Art of War*, May 14, 2013.

Silent Hawk, in order to meet the needs of the special forces seeking to counter enemy swarm attacks by motorcycle.[33]

[33] Kelsey D. Atherton, "Special Forces Are Getting a Stealth Motorcycle That's Silent and Deadly," *Popular Science*, May 26, 2017.

State of ~~the~~ what Field?: A Topic Modeling Approach to Assessing the Impact of Digital Issues on Terrorism Research

Christopher Whyte (L. Douglas Wilder School of Government and Public Affairs)

Abstract:

There is a pressing need to flesh out understanding of the contemporary research program on terrorism in the digital age. This article takes steps to do so by assessing the ideational landscape of such work. I employ a series of topic models on thousands of scholarly article abstracts to outline the thematic nature of such work, to assess the topical usage of cyber terror language in scholarly publications, and to map out the different legs of the research program. The results of this analysis show empirically that cyberterrorism is often cited as a qualifier but lacks thematic nuance and is tenuously linked to other major thematic topic areas.

Introduction

Over the past two decades, a host of scholarship in the security studies domain has problematized, conceptualized, and, where possible, examined cyberterrorism as a unique threat facing global society.[34] As any reader of such literature will quickly observe, however, there exists enduring division about cyberterrorism on several fronts.[35] Foremost

---

[34] For an early overview of the field of study see M. Conway, "What is Cyberterrorism? The Story so Far," *Journal of Information Warfare* 2, no. 2 (March 2003): 33–42. For a more recent account, see I. Awan, "Debating the Term Cyber-Terrorism: Issues and Problems," *Internet Journal of Criminology* 2045, no. 6743 (2014): 1–14.

[35] Awan, "Debating the Term," 2.

among these is fundamental definitional disagreement among scholars about what constitutes cyberterrorism itself. Does the term describe cyberwarfare (high-level disruptive operations enabled via the employment of malicious code) conducted by terrorists for coercive purposes?[36] Or is cyberterrorism simply constituted of actions taken by terrorists involving information and communications technologies (ICT), such as the mining and use of cryptocurrencies for nefarious purposes, the use of encryption to publish illicit content, or the use of social media to radicalize?[37] Though strong arguments can, and have, been made on both sides, there exists no clear consensus.

As one might expect given such a stark conceptual disagreement, scholars are also consistently split on the degree to which cyberterrorism actually threatens civil society and national security over and above the conventional shape of the terrorist enterprise. In recent work, Jervis, Lee, and Whiting (2016) array a series of potential cyberterrorist outcomes in noting that discourse invariably seems split between the sensationalist and the pedestrian.[38] A “cyber 9/11” or, as is often referenced more broadly in work on interstate cyber conflict, “cyber Pearl Harbor,” is often held up as a scenario realistic enough that scholars and practitioners should continue to prepare for an inevitable manifestation of cyberterrorism as massive disruption to infrastructure and society.[39] At the same time, a common

---

[36] See, for instance, D. Verton, “Black Ice: The Invisible Threat of Cyber-Terrorism,” New York: McGrawHill Osborne, 2003; and M. Pollitt, “Cyberterrorism-Fact or Fancy?” (2001). Accessed at http://www.csgeorgetown .edu/~denning/infosec/pollitt.html on June 21, 2017.

[37] E.g, M. Kenney, “Cyber-Terrorism in a Post-Stuxnet World,” *Orbis*, 59, no. 1 (2001): 111–28; J. Bronskill, “CSIS on Alert for Cyber Saboteurs: Spy Agency Monitors Threat to Computer Networks,” *Ottawa Citizen* (2001): 3; G. Weimann, “The Sum of all Fears?” *Studies in Conflict and Terrorism*, 129, no. 135 (2005); G. Weimann, *Terror on the Internet: The New Arena, the New Challenges*, US Institute of Peace Press, 2006; and G. Weimann, “Al-Qaida's Extensive Use of the Internet,” *CTC Sentinel (published by the Combating Terrorism C, US Military Academy at West Point)* 1, no. 2 (2008).

[38] L. Jarvis, L. Nouri, and A. Whiting, “Understanding, Locating, and Constructing Cyberterrorism,” in *Cyberterrorism*, ed. T. Chen et al., 25–41, NY: Springer, 2014.

[39] For work in this vein focusing on the development and impact of cyber threat alarmism, see S. Lawson, “Beyond Cyber-Doom: Assessing the Limits of

assessment is that guns and bombs are cheaper and, at least situationally, more effective than their cyber equivalents. If cyberterrorism is problematic, it is so partly because of the way that ICTs enable traditional terrorist operations and partly because of the way society perceives such an aid to political violence as a uniquely virulent problem set.

Furthermore, scholarship on violent extremism and the web is also characterized by disagreement on the degree to which ICTs themselves enable radicalization, subversion, and recruitment by terrorist outfits. Do new global information substrates and tools enable new modes of radicalism and new formats of ideational diffusion among extremist communities?[40] Or are digital tools simply evolutionary steps forward in the modes of communication and logistical support always found at the operational heart of the terrorist enterprise?[41]

Given the degree to which scholarship focused on cyberterrorism is centered on existential and definitional debate, identification of distinct substantive research trajectories can be a difficult task. After all, prospective threats are often the most difficult to problematize. Moreover, a multitude of political activities and developments sit at the intersection of the information revolution and the affairs of contentious nonstate actors on the world stage, a fact that presently fuels definitional ruminations more than it does empirical analyses. Here, I take steps to assess the ideational landscape of work focusing on the terrorist enterprise in the digital age.

Hypothetical Scenarios in the Framing of Cyber-Threats," *Journal of Information Technology & Politics* 10, no. 1 (2013): 86–103; and M. D. Cavelty, "From Cyber-Bombs to Political Fallout: Threat Representations with an Impact in the Cyber-Security Discourse," *International Studies Review* 15, no. 1 (2013): 105–22.

[40] E.g., A. Stenersen, "The Internet: A Virtual Training Camp?" *Terrorism and Political Violence* 20, no. 2 (2008): 215–33; and C. Edwards and L. Gribbon, "Pathways to Violent Extremism in the Digital Era," *The RUSI Journal* 158, no. 5 (2013): 40–47.

[41] See, e.g., D. C. Benson, "Why the Internet is Not Increasing Terrorism," *Security Studies* 23, no. 2 (2014): 293–328.; and W. McCants, "Testimony, US House of Representatives, Subcommittee on Counterterrorism and Intelligence, Jihadist Use of Social Media: How to Prevent Terrorism and Preserve Innovation, 6 December 2011," Retrieved Sept. 21, 2014, homeland. house. gov/sites/ homeland. house. gov/files/Testimony% 20McCants. pdf.

Where recent efforts on expanding the field both qualitatively and quantitatively have largely focused on how researchers should organize their efforts in a methodological sense, I focus on identifying discrete research trajectories apparent in existing scholarship. Though new veins of focus will undoubtedly characterize the landscape of cyberterrorism research over time, existing pathways suggest the clearest areas in which the diverse researchers who occupy this space might direct their energies.

In the sections that follow, I consider two primary questions. First, what themes constitute work that lexically centers on "cyberterrorism?" Second, what has the impact of the information revolution been on scholarship on extremism and political violence? I approach these questions in two primary ways. First, I obtain and assess topic models for scholarship taken from across numerous fields of academic study focused on "cyberterrorism" as a distinct concept. I then engage in an examination of both work focused on terrorism that distinctly engages with digital topics and research in the interdisciplinary cybersecurity field that covers terrorism and political violence. Here, I employ both traditional and dynamic topic modeling approaches to consider the impact of digital issues on the terrorism field, broadly construed. This effort speaks to a foundational disagreement still apparent in scholarly debates in the terrorism studies field on the nature of digital developments as either evolutionary or revolutionary.

The results of this analysis make several contributions to the literature on nonstate cyber conflict and extremists' use of ICT in world affairs. First, I demonstrate that cyberterrorism remains an indistinct concept that is nevertheless regularly employed in high-level descriptions of scholarship. Moreover, what thematic distinction exists in groupings that describe computer network security and risks to industry (and infrastructure) appears limited to specific academic fields, particularly computer science and business. These findings suggest that "cyberterrorism" is often employed as a qualifier in argumentation focused on semi-related topic matter. Moreover, while digital issues *do* appear as semantically distinct from other topics in the terrorism studies field, they manifest only when sufficiently expansive latent thematic model parameters are specified. The same is true of topics pertaining to terrorism studies found in the literature on cyber security and Internet governance, though such topics are more distinct than is the case with ICT-oriented work. Measures of topical similarity confirm that distinct semantic themes in this vein are referenced in small part across a wide swathe of research work but tend to have few major thematic connections to other topics, indicating thematically distinct engagement but limited impact on the high-level topical landscape of scholarship. Dynamic models that track topics over time support this assessment and demonstrate that the significance of digital keywords to the lexicon of core terrorism studies topic areas is a limited and recent phenomenon, likely triggered by increased focus on the Islamic State and other Islamist online organization over the past half decade.

The State of the Cyberterrorism Field

Research focused on the impact of the information revolution on the terrorist enterprise might be split into two categories: research seeking to problematize "cyberterrorism" as a discrete concept with related empirical manifestations and work that generally ignores such frameworks in assessing the use of ICTs by nonstate actors. Both are of interest to this study. The first of these categories covers a broad gamut of topics ranging from the highly conceptual and nonspecific to thick accounts of particular operations.[42] At the heart of the research program on cyberterrorism, however, is a simple fact: large-scale disruptive acts of terror enabled and executed *entirely* via the use of ICTs have not yet taken place.[43] To a degree, one might reasonably think that this fact limits the ammunition of those scholars who claim that "cyberterror" is entirely about such disruption.[44] If there is no evidence to support the notion that terrorist

[42] For an overview, see *inter alia* Weimann, *Terror on the Internet*; Chen et al., *Cyberterrorism*; L. Jarvis et al., "The Cyberterrorism Threat: Findings from a Survey of Researchers," *Studies in Conflict & Terrorism* 37, no. 1 (2014): 68–90; McCants, "Testimony" (2011); and C. Archetti, *Understanding Terrorism in the Age of Global Media: A Communication Approach*, NY: Springer, 2012.

[43] Weimann, "The Sum of All Fears?"

[44] A perspective first significantly outlined in a series of works by D. Denning circa 2001. See, for instance, Denning, "Activism, Hacktivism, and Cyberterrorism: The Internet as a Tool for Influencing Foreign Policy," *Networks and Netwars: The Future of Terror, Crime, and Militancy* 239 (2001): 288; and

organizations are set up to attempt coercion via cyberattack, then proponents of the idea that *all* digital elements of terrorist campaigns factor into the concept arguably boast a superior case for what should presently be considered to be "cyberterror."[45] And yet, it is not apparent that the absence of a defining "cyber 9/11"-style event is negatively indicative of the probability of such incidents taking place in the future. Indeed, if anything, work covering the increasing global expansion of a black market malware economy,[46] the development of organized criminal service operations[47] and the work of hackers linked with groups like Islamic State[48] suggests that coercive cyber assaults from terrorist sources will inevitably be a feature of the landscape of international affairs.[49] Whether such features will be exceptional or common is unclear.

Beyond debate over the real shape of cyberterrorism, much work in this vein-at least in the security studies fields, broadly defined-also focuses on antagonistic actions taken by terrorists online from a critical

---

"Cyberwarriors: Activists and Terrorists Turn to Cyberspace," *Harvard International Review* 23, no. 2 (2001): 70.

[45] Pollitt, "Cyberterrorism" ; also see Bronskill, "CSIS on Alert "and Gabriel Weimann, *Cyberterrorism: How Real is the Threat?* Vol. 31. United States Institute of Peace, 2004.

[46] See, among others, A. D. Romeo, "Hidden Threat: The Dark Web Surrounding Cyber Security," *N. Ky. L. Rev.* 43 (2016): 73; and M. Castelluccio, "The Silk Road on the Dark Web, *Strategic Finance* 99, no. 1 (2017): 55.

[47] E.g., C. Everett, "Ransomware: To Pay or Not to Pay?." *Computer Fraud & Security* 2016, no. 4 (2016): 8–12.

[48] See A. B. Atwan, *Islamic State: The Digital caliphate*, University of California Press, 2015.

[49] For work assessing the requirements and likeliness of cyber coercive operations, see D. Flemming and N. Rowe, "Cyber Coercion: Cyber Operations Short of Cyberwar," in *Proceedings of the 10th International Conference on Cyberwarfare and Security ICCWS-2015, Skukuza, South Africa,* March 2015, 95–101; C. Whyte, "Ending Cyber Coercion: Computer Network Attack, Exploitation and the Case of North Korea," *Comparative Strategy* (2016); and J. R. Lindsay and E. Gartzke, "Coercion through Cyberspace: The Stability-Instability Paradox Revisited," in *The Power to Hurt: Coercion in Theory and in Practice,* K. M. Greenhill and P. J. P. Krause, eds., New York: Oxford University Press, Forthcoming.

perspective.[50] The construction of the cyberterrorist threat is of significance for security researchers and practitioners for a number of reasons. Work in this area has principally noted the significance of a constructivist set of critiques of the phenomenon as related to better understanding of how impactful cyberterror might be as differentiated from more traditional elements of the terrorist enterprise. Since terrorism is inherently about coercion of social and political functions of a given national and transnational system, understanding of how cyberterrorism is framed inevitably lends itself to better comprehension of the motility of digital antagonism. In other words, threat construction matters a great deal because it can signify the nature and scope of sociopolitical responses to terrorism and because perceptions of threat constructs inform terrorist understandings of the relative value of different approaches.

In spite of some clear division in the field along definitional lines, some key contributors agree that "cyberterrorism" suffers as a distinct concept from a range of shortcomings. Foremost among these, as Jarvis and MacDonald note, is the fact that "cyberterrorism" is itself an extremely imprecise term that connotes a broad range of potential activities that don't map well onto the traditional parameters of research on the terrorist enterprise. In addition to the fact that several common terms (i.e., "cyberwar," "cyberjihad," "hactivism," etc.) are often used interchangeably with "cyberterrorism" in discussing general threats, more specific ruminations on techniques and tactics often gloss over the tenuous nature of the link between cyber actions and terrorist acts.[51] Certainly, the scenario of cyberattacks perpetrated by terrorist actors in order to realize some sort of political outcome would constitute cyberterrorism in the strictest sense of the term. But the use of ICTs more broadly to antagonize,

---

[50] E.g., M. Conway, "Media, Fear and the Hyperreal: The Construction of Cyberterrorism as the Ultimate Threat to Critical Infrastructures" (2008); L. Jarvis et al., "Constructing Cyberterrorism as a Security Threat: A Study of International News Media Coverage," *Perspectives on Terrorism* 9, no. 1 (2015); and L. Jarvis et al,. "Unpacking Cyberterrorism Discourse: Specificity, Status, and Scale in News Media Constructions of Threat," *European Journal of International Security* 2, no. 1 (2017): 64–87.

[51] Jarvis et al., "Constructing Cyberterrorism."

organize, and mobilize is not unique to the terrorist enterprise.[52] Even where terrorists may be more likely to employ ICTs for illicit disruptive and circumventive reasons than the average non-state actor (such as inciting adherents to violence with hate speech, soliciting funding from criminal organizations, or stealing demographic data from governments in order to plan attacks),[53] the same might be broadly true of a host of other contentious nonviolent participants in world affairs, from countercultural advocacy groups and cultist organizations to insurgent movements and loosely-defined hacker collectives.[54] Such antagonism is not necessarily linked with the defining feature of terrorism, i.e., acts of terror designed to coerce. The "cyberterrorism" field suffers, in short, from a lack of conceptual ownership of much of what constitutes the phenomenon the term allegedly describes.[55]

The analysis that follows is intended to help remedy this problem with the cyberterrorism field and contextualize scholarly efforts by mapping out distinct threads of the field's emerging research program. In doing so, it aims to augment broad advice for research practices in this vein with a more detailed read of how distinct thematic lines of approach manifest and where there exists empirical (i.e. topical) diversity within them. The empirical assessment of thematic trends in the field aims to answer three questions. First, is there clarity of conceptual and empirical nuance, even if divided, in work that focuses on "cyberterrorism?" Second, have information security topics had a semantically meaningful impact on the content of core elements of the terrorism and political violence field of study? And finally, what are the distinct legs of the research program on digital issues within the broader terrorism studies field? Answering these

---

[52] A point conceded by many scholars. See, among others, R. Heickerö, "Cyber Terrorism: Electronic Jihad." *Strategic Analysis* 38, no. 4 (2014): 554–65.
[53] For perhaps the best-known description of such activities in the aggregate, see G. Weimann, "Virtual Disputes: The Use of the Internet for Terrorist Debates," *Studies in Conflict & Terrorism* 29, no. 7 (2006): 623–39.
[54] For an introduction to such perspectives, see Denning, "Activism, Hacktivism," 288.
[55] M. Conway, "Determining the Role of the Internet in Violent Extremism and Terrorism: Six Suggestions for Progressing Research," *Studies in Conflict & Terrorism* (2016).

questions speaks directly to longstanding divisions in how scholars have approached the study of cyber terror, broadly writ, and is critical for the empirical expansion of the field called for in recent years by Conway, Jarvis and others.

Topic Modeling as an Approach

Traditionally, any attempt to systematically examine a field of scholarship runs into distinct measurement problems, from selection bias that flavors small-scale qualitative studies of the "most important" pieces of scholarship in a given vein to an inability to infer meaning from certain bibliometric data. Bibliometric attempts to uncover trends in research over time are particularly tempting for scholars, as both categorical and usage statistics commonly available at the article level ostensibly allow researchers to study the behavior of elements of the academy with regards to a specific topic. Among other things, such analyses can be useful for determining if there exist gender biases in citation practices, if institutional structure and imperatives prompt different modes of approach to publication of research on a topic, or if common funding discrepancies across academic fields lead to variably useful scholarly products.

As is often true of rating systems that do not derive directly from value intrinsically obvious in data, however, bibliometric approaches suffer from a poverty of meaningfulness of metrics.[56] Does an increased volume of work focusing on one particular topic over time really represent greater interest in the field or an expansion of perspectives being brought to the table? It may be the case that a field of study is undergoing a programmatic transformation of scholarship production expectations wherein research is being published in a more piecemeal fashion than was previously the case. Likewise, surging reference to a particular topic may stem from increased real-world focus on a subject and not a diversification of actual scholarly thought. The same may be said for direct usage of citation-derived metrics, where favor may be given to the "stars" of a given field or those at higher levels within professional cohorts.

---

[56] For a survey of such arguments, see C. W. Belter, "Bibliometric Indicators: Opportunities and Limits," *Journal of the Medical Library Association: JMLA* 103, no. 4 (2015): 219.

One major development in the computer science field-and specifically the machine learning field-to remedy such shortcomings has involved the construction of algorithmic tools for broad-scoped lexical analysis of sources that essentially produces meaningful data directly from text. One major set of tools is the topic model. Topic models are statistical models that consider collections of documents to be made up of common themes irrespective of the breakdown of documents themselves.[57] In essence, documents can be thought of as containing different themes and topics determined by the contours of the larger group of documents. Topic models assume that the broader corpus reflects a fixed vocabulary from which topics emerge. Distributions of topics are obtained simply via algorithmic assessment of term collocations and frequencies, with the result that individual documents are constituted of multiple themes that are only apparent at the higher corpus level.

Put more simply, probabilistic topic models allow for the latent discovery of thematic trends across an unstructured corpus of documents.[58] Topic models do not require prior input of parameters by researchers (though semi-structured models are not uncommon in research employing topic models), a fact that remedies issues of semantic bias introduced at the design phase that are common in the social sciences. Finding themes is an inductive process. Latent discovery of topics thus provides a unique capability for researchers to view the "ground truth" of trends in document construction. Thematic trends, existent across entire input corpuses but annotated as proportions across individual documents, are useful as an artifact for qualitative analysis *and* for quantitative investigations wherein semantically meaningful latent trends across a

---

[57] D. M. Blei et al., "Latent Dirichlet Allocation." *Journal of Machine Learning Research* 3, no. Jan (2003): 993–1022.

[58] For full details on the evolution of topic models, see T. L. Griffiths et al., "Hierarchical Topic Models and the Nested Chinese Restaurant Process," in *Advances in Neural Information Processing Systems*, 17–24, 2004; D. M. Blei and J. D. Lafferty, "A Correlated Topic Model of Science," *The Annals of Applied Statistics* (2007): 17–35; D. Ramage et al., "Partially Labeled Topic Models for Interpretable Text Mining," in *Proceedings of the 17th ACM SIGKDD International Conference on Knowledge Discovery and Data Mining*, ACM, 2011, 457–465; and Blei, "Probabilistic Topic Models," *Communications of the ACM* 55, no. 4 (2012): 77–84.

particular dataset can be referenced in coding of discrete sociopolitical phenomenon as easily as might researcher-determined variables (and without the requisite concern about input bias).[59]

Topic models can be obtained in either an unsupervised or semi-supervised fashion.[60] In other words, researchers *can* limit their inputs to setting parameters (topic frequency, smoothing, etc.) for topic generation without imparting any prior coding bias on the text-to-data process. But they might also identify existing divisions in a document corpus by annotating documents accordingly (i.e., using a control variable to force topic generation artificially around a pre-set descriptor variable, such as single vs. multiple authors, "high" citation count vs. "low," etc.).[61] In this way, scholars are given tools to assess the viability of other tools and data available for a given topic. The sections below make additional use of the ability to force topic generation with minimal researcher input by taking the results of unsupervised topic models, using latent themes detected across the corpus to assign topical values at the document level,[62] and then further obtaining topic models for what is a multi-layered study of topicality in the cyberterrorism field.

Data

In order to examine the shape of the dispersed field of scholarly works that cover a range of issues at the intersection of the information revolution and the terrorist enterprise, I employ the texts of nearly 5,000 peer-reviewed article abstracts drawn from across nearly three decades and downloaded alongside relevant bibliometric data from Web of Science. The sections below assess the field in several different slices, from research that focuses centrally on the term "cyberterrorism" itself to the broader landscape of research into both terrorism and cyber security.

---

[59] In this way, topic models are more than simply a unique tool for general observation of trends in text distributions. They are a way to control for the traditional input biases that stand to skew the conveyance and results of research wherein researchers systematically adopt a subjective and potentially not replicable coding scheme.

[60] Blei, "Probabilistic."

[61] Ibid.

[62] As in Ramage et al., "Partially Labeled Topic Models."

Latter sections address the combined corpus and employ the semi-supervised design outlined above to produce a multi-layered map of existing arms of research on issues of digital age terrorism.

Source data was collected from Web of Science using a series of keyword collocation search inputs aimed at capturing all conjugations and variations of terminology related to cyberterrorism. Initially, this included specific reference to the term "cyberterrorism" and variants. Beyond this initial focus on the core term, however, the aim was to collect all relevant scholarly publications across the two main fields of study within which discussion of cyberterrorism is most likely to appear, namely, terrorist studies and cyber security and governance research. In both instances, keyword collocation formulae were employed in order to capture all relevant source data (i.e., collocation of terms like cyber* within n words of defense, military, etc. and the additional provision of unique keywords like radical* or governance within *n* words of cyber*/Internet/digital, etc.) and specific unique terms (e.g., cyberjihad) were included to ensure complete coverage.

The choice to use abstracts as the primary input material follows precedent in a range of works in political science, business, computer science, and machine learning that have employed topic models in research.[63] Abstracts solve a challenge for machine learning algorithms that attempt to understand latent patterns in large amounts of input information wherein massive documents are too noisy for effective treatment. Not only is publication length variable across academic journals and other publications, but authors exhibit distinct writing styles and will diverge radically from their peers in how they reference relevant topics, establish theoretical foundations, and outline arguments. The abstract not only allows for a concise application of algorithms to source input information; it is also the element of any publication where authors must

---

[63] Abstracts are a common choice for such lexical analyses of scholarly bodies of work, as in J. Chang and D. M. Blei, "Relational Topic Models for Document Networks," in *International Conference on Artificial Intelligence and Statistics*, 2009, 81–88; A. J.-B. Chaney and D. M. Blei. "Visualizing Topic Models," in *ICWSM*, 2012; and J. D. Mcauliffe and D. M. Blei, "Supervised Topic Models," in *Advances in Neural Information Processing Systems*, 2008, 121–28.

summarize their premises, research questions, and arguments in a concise and accessible fashion.

Finally, researchers employing a topic modeling approach must select three sets of parameters for model output-the number of topics desired and a pair of smoothing parameters. In the sections below, various topic models are presented.[64] For the most part, the size of each (i.e., the number of topics) is the result of trial-and-error on the part of the researcher and validators to identify semantically meaningful proportions among the model outputs. Several different topic model sizes were chosen for presentation from a large number of produced models. The exception to this approach lies with the dynamic topic presented below. There, I obtain a Word2Vec model as a means for obtaining window topics without researcher input.[65] In essence, the number of topics is determined via an initial algorithmic bag-of-words assessment, a result that both informs the resultant topic model and is interesting as a measure of field diversity in itself. Smoothing terms-both term and topic terms-affect the granularity of word assignment to different topics. After experimentation and in line with other studies, I set both to 0.01.

What is Cyberterrorism?

What constitutes "cyberterrorism?" This question lies at the heart of much scholarly debate and investigation, particularly in those predominant elements of the terrorism studies community that hail from the political science, security studies, and psychology fields. The first task I turn to is a direct assessment of work that employs the term (and variants thereof, like "cyber-terrorism," "cyberterror," "cyber terror," etc.) itself. This is done, as described above, through application of a keyword, key

---

[64] For further details on the nature of smoothing and how choices might be made by the researcher, see A. Asuncion et al., "On Smoothing and Inference for Topic Models," in *Proceedings of the Twenty-Fifth Conference on Uncertainty in Artificial Intelligence*, AUAI Press, 2009, 27–34. For a description of selection in social science work, see S. Kaplan and Keyvan Vakili, "Studying Breakthrough Innovations Using Topic Modeling: A Test Using Nanotechnology Patents," 2013.

[65] As in D. Greene and J. P. Cross, "Exploring the Political Agenda of the European Parliament Using a Dynamic Topic Modeling Approach," *Political Analysis* 25, no. 1 (2017): 77–94.

term, and keyword collocation formula to all peer-reviewed publications available through the Web of Science research interface. The result is a corpus of around 500 articles that prominently reference or employ the term and its variants. A series of parameters for topic modeling via latent dirichlet allocation (LDA) were then experimented with such that a semantically meaningful breakdown of latent themes in such literature is represented. Figure 1 below displays initial results for a 12-topic model.

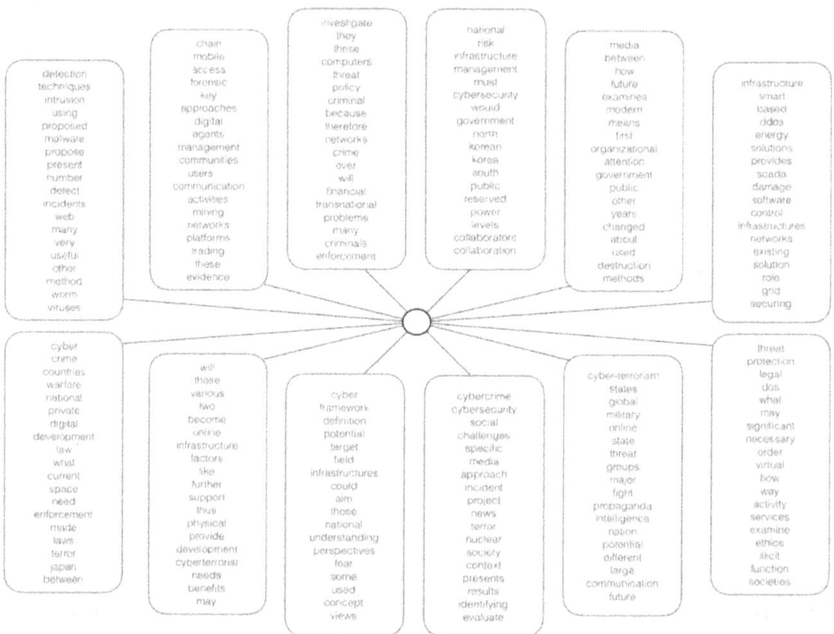

Figure 1. Results for a 12-topic LDA model obtained from nearly 500 article abstracts containing language" based around the term "cyberterrorism" and its variants.

The topics (the combinations of words grouped under each box) arrayed in Figure 1 contain several distinguishing features. In the top left, two topics clearly describe discrete thematic areas in technical cyber security. One, prominently grouping words like detection, malware, intrusion, worm and viruses together, clearly describes work focused on network security and the security of networked computers. The other, with strong topical relationship between words like chain, mobile, management, users, and platform, appears to describe the "lower" layers of computer security pertaining to the design of devices, the security of operating systems, and general issues in access control. Elsewhere, several topics make distinct reference to different threat modes or topic areas commonly differentiated in work on cyber security and cyberterrorism, including cyber threats to critical infrastructure (top right), cybercrime and legal responses (bottom left) and notable global hotspots of cyber tensions (top center).

However, the story of these model results is nevertheless one of vagueness about the "field" of cyberterrorism. Though there exist a few highly specific topics in the display in Figure 1, most are imprecise in their description of discrete themes. One possible explanation for this is that work containing reference to "cyberterrorism" is itself inherently multidisciplinary in nature. After all, the topics to be discussed under the cyberterror moniker (from actual computer/network security issues to infrastructure protection practices, law enforcement policies, and more) are likely variably of interest to scholars across different fields of academic study. Thus, in order to consider latent trends in work on cyberterrorism in greater context, Figure 2 breaks the corpus into different identifiable fields of academic research based on Web of Science categorizations and re-produces 6-topic models via LDA.

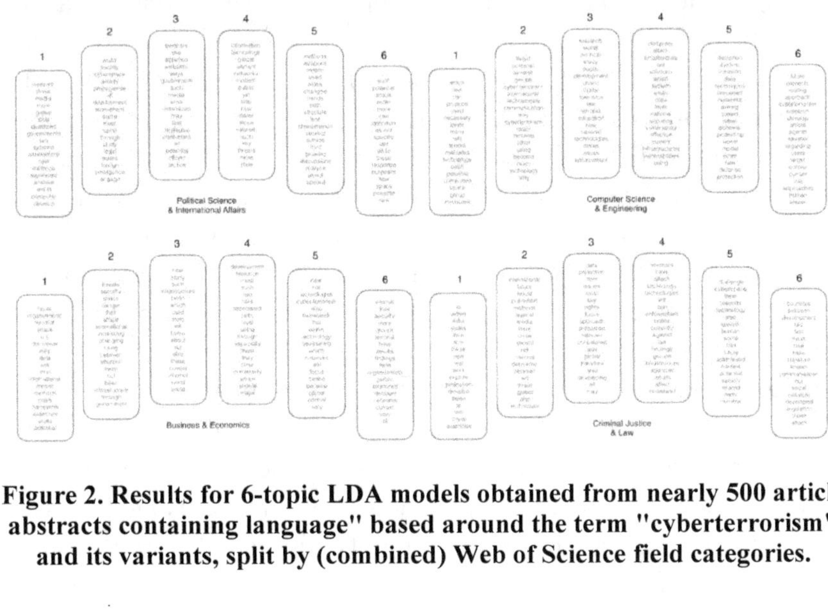

Figure 2. Results for 6-topic LDA models obtained from nearly 500 article abstracts containing language" based around the term "cyberterrorism" and its variants, split by (combined) Web of Science field categories.

Figure 2 arrays four 6-topic models by different academic fields of study. These are (1) political science and international affairs, (2) computer science and engineering, (3) business and economies, and (4) criminal justice and legal studies. Six topics were chosen following a series of experimental model production runs that suggested expanded models exhibited diminished semantic clustering. This is not unexpected, given that any categorical narrowing of focus of a given corpus usually reduces the vocabulary parameters with which the algorithm will work.

What was unexpected is the degree to which the granularity of a limited number of topics in the Figure 1 model appears to diffuse across fields of academic research. Technical topics that address the content of security concerns regarding terrorist involvement in computer network attack *are* more readily apparent in work in the computer science and engineering fields of study. There, however, remains a single topic devoted to such relatively more granular discussion of the components of a computer network attack threat (Topic 5). Others variably and vaguely group together based on common terms linked with the threat of cyberterrorism, from *attack, malicious* and *target* to *vulnerability, infrastructure, and digital.*

Likewise, research referencing cyberterrorism in other fields of study presents vaguely across the topics produced. Latent thematic similarity in most cases is limited to apparent descriptions of a threat (generally phrased), underlying global and societal developments, and the need for research itself. Though some topics, such as those in the business and economics field that contain regular use of terms like risk and protection, do suggest a linkage between the content of research and the unique imperatives of a given field of academic study, almost no topics are granular enough to be considered as distinct from those around it.

This initial result is troubling, though perhaps not particularly unexpected. As authors like Jarvis and MacDonald have noted in recent work attempting to problematize cyberterrorism, there exists little in the way of consensus among scholars about what the term connotes. Moreover, it is common practice for academic work referencing the concept to treat the term as synonymous with others, like cyberjihad or even cyberwar. The models above supports this read of work focusing on "cyberterrorism" itself as being distinctly indistinct. Indeed, insofar as the few more relatively granular themes in Figure 1 are not present across numerous identified topics, it seems reasonable to suggest that the term "cyberterrorism" and its variants are regularly employed to connote a broad-scoped threat potentiality and a distinct (though only at the highest possible level) cyber conflict concept referenced in yet-foundational discussions of the changing shape of security challenges in the digital age.

At the Intersection of Cyber and Terror

Naturally, the initial models discussed above do little to engage with the thematic granularity of work at the intersection of ICT and the terrorist enterprise. Rather, they focus on work that specifically contains the terminology of "cyberterrorism." Even insofar as the results above appear to qualitatively describe the vague condition of engagement with the concept of cyberterrorism, they cannot effectively describe the impact of the information revolution on work covering political violence, extremism, and terror in conflict. This section undertakes that task with a far more broad-scoped investigation of latent thematic trends in two sets of literatures: on (1) cyber security and Internet governance, and (2) terrorism, radicalization, and political violence. The primary aim here is to assess the degree to which ICT-specific content matter presents in the terrorism studies literature (and vice versa in the cyber/governance literature) in order that we might obtain a better understanding of how fully scholars have attempted to develop and diversify our understanding of digital age extremism. This is done in two ways: via (1) initial LDA assessments of the thematic scope of each field alongside similarness assessment of relevant topics and (2) use of non-negative factorized matric (NMF)[66] dynamic topic modeling to "view" the year-to-year evolution of scholarship focused on cyber issues in a combined corpus of both fields.

---

[66] See D. D. Lee and H. S. Seung, "Learning the Parts of Objects by Non-Negative Matrix Factorization," *Nature* 401, no. 6755 (1999): 788–91.

Cyberspace and Information Technologies within the Terrorism Literature

To what degree does content focused on ICT-related matters constitute a clear and distinct part of the terrorism studies field of study? To some degree, this might seem like a fool's errand insofar as ICT issues are almost inherently crosscutting. Discussion of a broad range of topics in the field might include the role of the Internet, the use of malware, or the part played by social media in enabling extremist activities. Nevertheless, topic modeling approaches have demonstrably been able to differentiate between the use of content in bodies of text as either pretext or thematically distinct and complex elements of literature. Table 1 and Figure 3 below demonstrate exactly this point.

Topic	Keys
1	*nuclear, weapons, will, threat, north, would, korea, mass, destruction, united, could, terrorists, u.s, against, such, use, proliferation, there, should, may*
2	*intelligence, been, crime, counter-terrorism, new, national, global, threats, counterterrorism, measures, such, community, policy, governance, threat, organized, through, cooperation, european, efforts*
3	*law, rights, human, against, legal, use, military, force, humanitarian, such, torture, under, freedom, criminal, armed, justice, right, protection, united, responsibility*
4	*suicide, social, organizations, violent, group, terrorists, radicalization, religious, movement, organization, how, study, organizational, analysis, members, ideology, individuals, case, extremism, who*
5	*september, after, bush, eleven, american, administration, president, 2001, nine, policy, years, iraq, since, new, over, afghanistan, george, were, had, taliban*
6	*data, study, countries, using, analysis, economic, domestic, results, civil, transnational, conflict, incidents, effect, find, than, effects, democracies, empirical, our, evidence*
7	*terrorists, they, can, government, when, may, counterterrorism, strategic, than, support, against, model, military, conflict, actors, terror, strategy, likely, will, use*
8	*islamic, qaeda, iraq, muslim, islamist, islam, al-qaeda, pakistan, western, against, syria, global, religious, jihadist, world, afghanistan, militant, radical, strategy, arab*
9	*policy, world, relations, regional, european, countries, economic, new, global, cooperation, asia, peace, east, foreign, development, region, europe, china, power, military*
10	*public, media, were, support, threat, study, news, about, citizens, how, survey, attitudes, policy, social, findings, perceptions, among, national, who, perceived*
11	*risk, information, internet, use, health, attack, such, biological, can, used, response, management, potential, cyber, technology, emergency, preparedness, communication, threats, training*
12	*how, research, analysis, studies, what, understanding, within, politics, discourse, about, can, critical, policy, been, argues, terror, paper, first, approach, through*

Table 1. Results for 12-topic LDA model obtained from 4,166 articles constituting the body of work in the Terrorism Studies field.

Table 1 displays the results for a 12-topic LDA model obtained from 4,166 article abstract taken from the Web of Science database. As with the previous section, articles were identified via reference to a unique formula of search keywords, key terms, and term collocations. Distinct from the models arrayed in Figures 1 and 2, there are clear high-level themes in this model, related to distinct functional and substantive elements of the research program on terrorism and political violence. In particular, there are several functional sub-elements of the field that describe the methodological topology of the field (Topics 6 and 12), others that contain key concept groupings (Topics 4 and 5), and yet others that relate to significant substantive organizers (Topics 1, 2, 3, 8, and 9). With regards to those substantive organizers, there are distinct high-level topics centered on global jihad, humanitarianism and rule of law in conflict, operational counterterrorism, and macro threats to international security. This last set of topics (particularly Topics 1 and 9) demonstrate the consistent engagement of the broader security studies community with those focused on extremism and political violence, either via reference to terrorism in broader security assessments or discussion thereof in the context of evolving global security conditions.

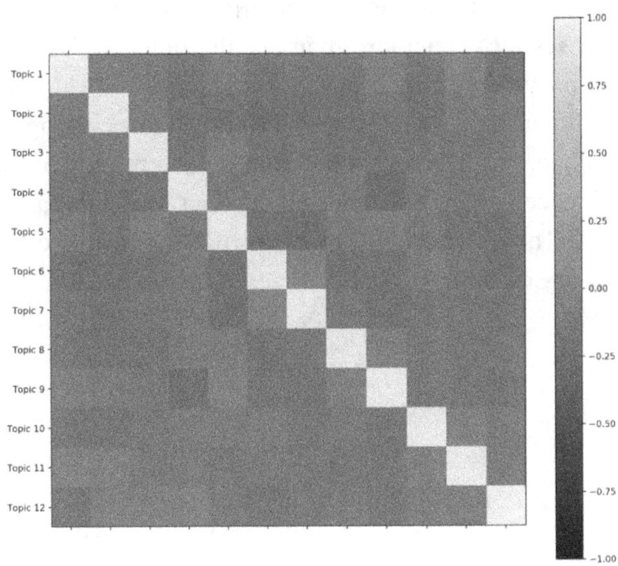

Figure 3. Correlation matrix visualization for Table 1's 12-topic LDA model.

Though keywords linked with ICT *do* appear prominently in Table 1's high-level modeling of the terrorism studies field (i.e., Topic 11's cyber, internet, technology, information, etc.), they appear alongside unrelated key terms like biological, health, and preparedness. This suggests that cyber or ICT topics remain indistinct in such high-level topical models of the field and are instead a significant element of a clear "new threats" theme. Interestingly, however, whereas one might be tempted to think that such "new threats" might be a regular feature in work studying extremism and political violence (i.e., a throwaway qualifier applied referentially in work more distinct at the field's highest levels), this appears not to be the case. Figure 3 visually displays Pearson's r correlation values for topic proportion similarities across the 12 topics produced in Table 1's LDA. With topic models, the general expectation is that low yield topic parameters (i.e., small requested numbers of topics to be returned) with appropriate smoothing parameter settings will return more thematically distinct topics than will more expansive ones. Figure 3 demonstrates this in showing the remarkably low, consistent r values associated with each topic pairing. Of specific interest here is the fact that Topic 11 is not notably "stickier" than are the other topics found. This would seem to provide evidence in favor of the null for any hypothesis assuming that focus on "new threats" means general engagement and qualification alone. At least at this high level, there appears to be reasonable evidence of the existence of distinct "new threat" topics in work in the field. Table 2 and Figure 4 below pick up further on this point.

Topic	Keys
1	*will, they, should, can, would, may, than, when, must, even, power, because, such, terrorists, all, many, those, them, likely, some*
2	*crime, economic, been, new, organized, trade, global, governance, challenges, maritime, transnational, development, attention, changes, institutions, issues, process, since, century, piracy*
3	*rights, human, law, legal, against, use, such, humanitarian, under, laws, torture, responsibility, force, right, intervention, norms, protect, freedom, detention, case*
4	*threats, law, system, actors, enforcement, order, new, global, development, regime, traditional, different, democratic, such, environmental, within, non-state, through, democracy, legal*
5	*pakistan, iraq, syria, afghanistan, iran, party, india, turkey, indian, taliban, islamic, against, parties, been, regional, country, pakistani, government, support, turkish*
6	*was, were, after, had, september, during, new, years, over, been, 2001, events, 11-Sep, since, first, time, period, attack, two, there*
7	*intelligence, counterterrorism, policy, counter-terrorism, been, government, community, police, measures, agencies, national, policies, prevent, strategy, agency, such, role, how, level, implementation*
8	*conflict, conflicts, armed, civilians, bin, insurgency, against, boko, african, killing, laden, haram, africa, insurgents, peace, non-state, nigeria, such, rebel, forces*
9	*conflict, conflicts, armed, civilians, bin, insurgency, against, boko, african, killing, laden, haram, africa, insurgents, peace, non-state, nigeria, such, rebel, forces*
10	*support, threat, public, survey, citizens, attitudes, study, perceptions, were, trust, opinion, toward, among, perceived, who, results, national, respondents, effect, americans*
11	*risk, health, public, response, emergency, preparedness, disaster, biological, risks, training, events, attack, resilience, management, event, bioterrorism, such, needs, natural, community*
12	*politics, discourse, through, how, global, critical, within, practices, identity, terror, relations, explores, narratives, power, argues, such, ways, argue, theory, narrative*
13	*economic, countries, data, transnational, study, effect, country, incidents, domestic, impact, results, using, analysis, electoral, show, growth, find, number, period, our*
14	*model, behavior, ireland, analysis, northern, models, study, present, republican, personal, activity, well, implications, specific, using, response, most, discussed, proposed, results*
15	*civil, than, democracies, likely, data, study, effects, using, democratic, countries, research, domestic, empirical, other, regime, argue, results, find, less, use*
16	*what, concept, first, about, question, second, one, other, how, warfare, terror, argues, new, contemporary, been, debate, his, both, essay, paper*
17	*military, operations, forces, strategic, force, against, use, strategy, capabilities, defense, efforts, effective, civilian, threats, nato, campaign, role, afghanistan, while, order*
18	*terrorists, terror, attack, government, target, targets, may, model, when, can, policies, counterterrorism, deterrence, actions, governments, strategic, they, against, potential, such*
19	*social, crisis, media, people, society, internet, become, world, crises, other, how, urban, public, new, one, can, action, all, communication, country*
20	*was, had, his, they, after, french, russian, movement, organization, eta, left, end, france, party, government, basque, revolutionary, terrorists, been, what*
21	*media, how, analysis, news, coverage, about, study, role, events, framing, public, official, were, discourse, examines, information, issues, rhetoric, paper, content*
22	*qaeda, network, islamic, jihadist, global, iraq, threat, networks, movement, group, al-qaeda, qaeda's, lone, ideology, isis, jihad, fighters, organization, structure, united*
23	*suicide, israel, women, palestinian, israeli, conflict, gender, female, hamas, bombing, women's, terrorists, hezbollah, lebanon, bombers, palestinians, peace, motivations, bombings, gaza*
24	*nuclear, weapons, destruction, mass, threat, proliferation, chemical, biological, wmd, could, use, terrorists, united, will, would, global, potential, such, materials, weapon*
25	*research, studies, analysis, literature, study, understanding, radicalization, theoretical, approach, case, empirical, how, framework, field, social, work, academic, our, some, into*
26	*religious, islamic, muslim, islam, islamist, religion, radical, world, muslims, western, arab, violent, identity, extremism, middle, movements, cultural, militant, among, east*
27	*organizations, group, organizational, violent, organization, they, factors, social, ideology, members, why, support, militant, goals, study, use, how, individuals, can, terrorists*
28	*policy, u.s, united, american, foreign, bush, administration, president, september, iraq, 2001, power, strategy, obama, his, policies, new, against, administration's, domestic*
29	*information, financial, technology, cyber, paper, system, infrastructure, can, money, will, use, activities, technologies, national, internet, new, used, systems, water, data*
30	*north, korea, china, law, korean, relations, chinese, sea, japan, rights, south, edited, human, diplomatic, united, u.s, diplomacy, korea's, china's, america*

Table 2. Results for 30-topic LDA model obtained from 4,166 articles constituting the body of work in the Terrorism Studies field.

Table 2 displays thirty topics produced via LDA from the corpus of terrorism studies scholarship. Thirty topics were chosen for several reasons. First, in experimentation the highest proportion of semantically meaningful themes was evident in models producing between twenty-five and thirty-seven topics. Second, and perhaps more importantly, this production of thirty topics represents the smallest number of topics wherein there is a clear "digital issues" subgrouping distinct from other terms. Topic models that include higher numbers of topics function differently than do those featuring small topics. That is not to say that they are produced via a different process or that the algorithm involved actively alters its treatment of variables. Rather, it is simply the case that topic models specifying higher numbers of topics to be produced feature greater stratification of significant topics wherein the first few specified enduringly describe themes that uniquely define documents in the corpus and later topics more truly describe distinct sub-themes that appear in (but do not singularly define) documents. Though the input challenge for the researcher is simply to produce topic models that maximize semantic meaningfulness among topics produced, different output formats allow for more or less granular assessments of the ideational construction of (in this case) an academic field.

Much as was the case with the 12-topic high level model of the terrorism studies literature, the topics displayed in Table 2 largely retain clear semantic meaning when considered alongside one another. Aside from a couple of functional topics (i.e., those dominated by common functional words not caught by a stop list in pre-processing, usually unique to a corpus), most topics describe the methodological topology of the field, key concept groupings, and significant substantive organizers. Naturally, these topics are more diverse in this expanded assessment of lexical tendencies in the terrorism studies field. Whereas Table 1 outlined a "new threats" topic that include cyber and ICT keywords among unrelated ones, Table 2 displays a more nuanced read of the field wherein issues like biological threats to national security and cyber security are fleshed out. That said, the "digital issues" topic (Topic 28) is only distinct in this expanded model. The question, as the results above suggest, is whether ICT focus in research presents as a unique program of study or a regular reference point that retains semantic distinction in terminological terms only (i.e., the wording is unique, but there is no distinct scholarly engagement).

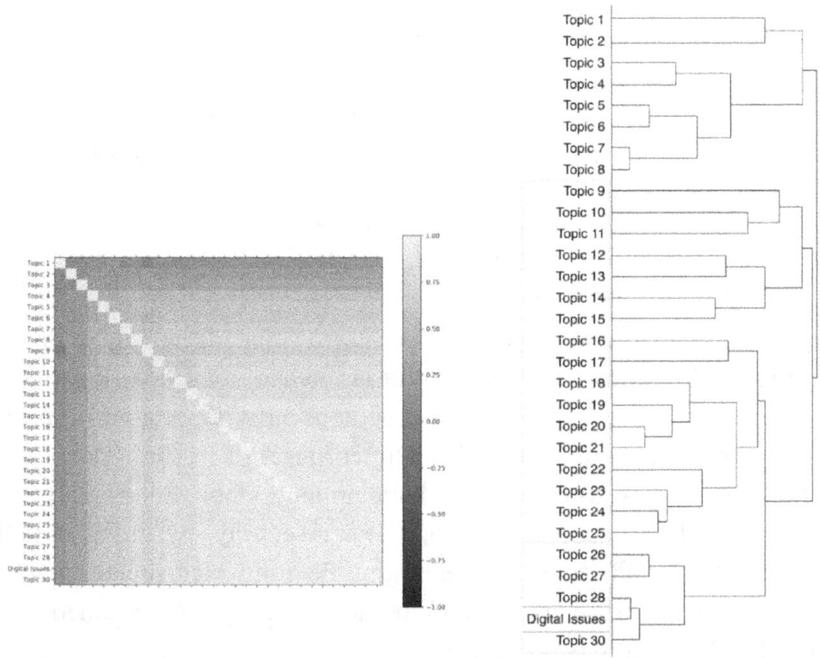

Figure 4. Correlation matrix and dendrogram visualization for Table 2's 30-topic LDA model.

The visualizations displayed in Figure 4 above address this point in two ways. On the left is displayed a correlation matrix visualization for Table 2's 30-topic LDA model. Where the relatively low *r* scores illustrated in Figure 3 above are common for models wherein a low number of topics is specified for return, more expansive topic models should display a clear spectrum of topic correlation where initial topics reflect distinct semantic themes in a given corpus (such as dominating paradigms in an academic field of study) and lower ranked topics reflect distinct substance below the level of overarching themes. Figure 4 does just this, with functional and major substantive topics at a certain height exhibiting higher correlation with other topics at similar heights. In the dendrogram above, the sole "digital issues" topic is surprisingly divorced in similarity from other topics, suggesting that the topic is distinct but not thematically close to many others. At the same time, the correlation matrix (reflecting Pearson's *r* scores opposite the similarity matrix used to produce the clustering visualization) suggests that Topic 29 co-occurs regularly with all but the highest-numbered topics. The narrative that seems to emerge from these figures is, thus, that "digital issues" are a distinct sub-theme of the terrorism studies literature that are referenced in small part across a wide swathe of scholarly works but that have few major thematic connections to other topics.

Terror in the Literature on Cyber Conflict and Security

Flipping the prompt on its head, to what degree do topics in terrorism and political violence present in the body of scholarly work on cyber conflict and security? Here, it is perhaps most important to consider the scholarly communities that different corpuses of research documents (in this case, abstracts) proxy for. In attempting to look at terrorism topic matter in the literature on cyber conflict and security, we are essentially looking at the broad scope of an emerging field in which social scientists (primarily political scientists and related security policy-practitioner researchers) are actively self-organizing in line with emerging substantive frameworks. This stands in relatively stark contrast with literature in the terrorism studies field where the above analysis is essentially considering a more substantively focused family of research communities. With both bodies of work, of course, the primary question is of impact of digital topics on the broader field such that we might adjudicate on the notion that ICTs have meaningfully altered the shape of different research programs.

Topic	Keys
1	*social, use, terrorist, online, groups, public, activities, they, communication, digital, media, world, who, people, through, organizations, terrorists, science, been, society*
2	*terrorism, war, state, strategic, global, world, digital, what, one, other, terrorist, china, organizations, foreign, current, united, military, become, how, also*
3	*threat, attacks, over, would, nuclear, world, most, while, power, threats, risk, military, against, technology, into, major, may, governments, russia, time*
4	*operations, military, defense, war, cyberspace, network, force, use, capabilities, than, offensive, model, but, attack, cyberwar, terms, been, work, order, computer*
5	*system, systems, data, attacks, attack, model, against, services, infrastructure, management, network, risk, service, our, based, detection, using, critical, threats, access*
6	*national, critical, government, challenges, space, all, infrastructure, countries, development, cybersecurity, threats, strategy, strategies, measures, global, nato, defence, leadership, must, needs*
7	*parties, media, between, websites, party, was, participation, campaign, two, both, study, were, social, than, democracy, using, news, data, however, public*
8	*governance, policy, power, conflict, cyberspace, development, how, actors, but, government, society, smart, between, theory, public, cyber-security, been, politics, city, complex*
9	*how, use, malware, they, there, attacks, used, software, but, also, two, research, analyze, level, most, between, tools, both, make, problem*
10	*law, human, rights, convention, protection, treaty, china, criminal, against, space, weapons, act, outer, sea, arms, jurisdiction, humanitarian, armed, territorial, laws*
11	*legal, but, law, state, electronic, north, there, privacy, including, technologies, data, private, cyberspace, should, korea, networks, traditional, some, most, south*
12	*cyberspace, research, theory, analysis, literature, cyberterrorism, policy, making, been, issues, decision, based, questions, there, then, threat, between, study, theoretical, national*

Table 3. Results for 12-topic LDA model obtained from nearly 600 articles constituting the body of work in the cyber security and conflict field.

Table 3 arrays a dozen topics obtained from a corpus of nearly 600 articles constituting the body of work in the cyber security and conflict field. As before, articles were identified via reference to a unique formula of search keywords, key terms, and term collocations. Interestingly, while the literature on cyber conflict presents as a distinct topology of research areas, several identifiable thematic threads appear to be constituted of general threat language more than of key terms one might expect. Though the table above reports only the top ~15 keywords associated with each topic, this trend holds through the top 40 keywords presented for each topic in the LDA production process.

Among the topics, there are several references to terrorism. However, while topics containing relevant keywords have distinguishing characteristics in relation to others, terrorism-inclusive topics are among those that contain the most generic terminology. While Topic 1 generally suggests the shape of research focused on terrorists' use of the web as an organizing and mobilizing tool, Topics 2 and 12 would be difficult to characterize beyond a label like "cyber threat terminology." By contrast, other areas of the field topically present as relatively distinct. Certainly there are other general terminology topics, but Topic 5 suggests the shape of work on threat management related to infrastructural threats, Topic 7 appears to describe scholarship on the democratic information environment, Topic 10 distinctly describes work on international law, and Topic 11 seems to emerge from analyses specifically focused on the Korean Peninsula.

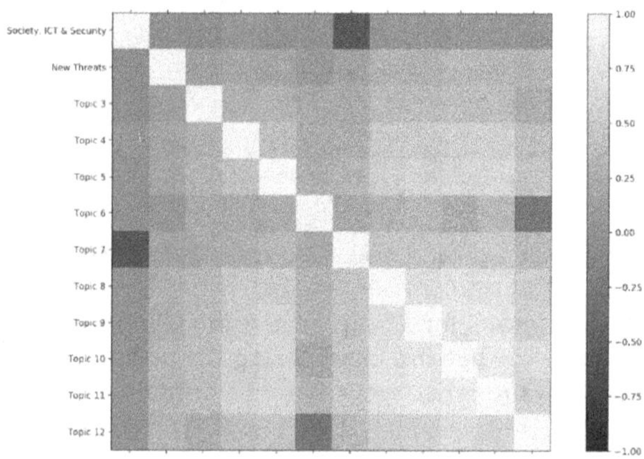

Figure 5. Correlation matrix visualization for Table 3's 12-topic LDA model.

Much as was the case with Table 1's initial high-level topic modeling of terrorism studies literature, Table 3's results suggest that terrorism is an indistinct element of the cyber conflict field employed alongside other traditional threat nomenclature in scholarly works. Figure 5 above considers the similarity of the 12 initial topics. The results are quite interesting. Topic 1, perhaps the clearest and most distinct terrorism-oriented topic, presents as analytically distinct from other areas of the cyber conflict field. Again, as noted above, it is to be expected that lower-numbered topics will correlate only minimally with those that follow and that higher-numbered ones will likely co-occur increasingly strongly with other topics. This is largely to do with the way in which topic model outputs present themselves in the initial identification of thematically distinct major themes in a given corpus that then give way to, depending on the complexity of the text in the corpus, either less distinct major themes or minor themes that present as increasingly present across research in small, distinct amounts.

In Figure 5, Topic 1 is extremely dissimilar to other topics. This is potentially because Topic 1 describes the shape of one of the most relatively distinct program of study that exists in the cyber conflict literature: that of terrorists' use of the web to radicalize and organize. One exception to this is a high negative r score with respect to Topic 7. This is perhaps unsurprising if we assume that Topic 7, in dealing with democratic processes and political elements of democratic information environments, significantly contains work aimed at assessing e-governance and cyber-augmented information operations. Nevertheless, it is curious that there appears to be little crossover in two domains so concerned with the public sphere.

Topic	Keys
1	*communication, research, people, been, network, use, results, study, students, social, groups, they, being, become, studies, who, one, there, university, than*
2	*state, global, digital, one, but, all, european, sovereignty, within, threats, world, very, effective, system, systems, cyberspace, only, steps, hand, alliance*
3	*risk, threats, governments, framework, response, attacks, covert, tools, address, events, organizations, action, responses, disruptive, proposed, each, cost, industry, into, our*
4	*system, model, cyberspace, models, view, how, defense, deterrence, complex, work, modern, possible, concept, cyber-attack, assurance, better, make, deception, point, rather*
5	*government, critical, services, infrastructure, model, e-government, public, private, websites, sector, national, activities, implementation, well, ict, especially, protection, business, administration, role*
6	*space, under, systems, documents, performance, use, reports, principles, management, elements, all, strategy, control, european, activities, national, current, rights, measures, union*
7	*parties, party, campaign, small, media, major, election, minor, politics, chapter, than, both, news, canadian, within, web, competition, conservative, support, local*
8	*policy, strategies, cyber-security, research, discourse, how, social, practices, explores, both, argues, but, leadership, attempt, understanding, strategy, were, actors, meaning, safe*
9	*norms, efforts, actors, private, issues, should, national, attacks, approach, while, cyberspace, issue, they, deterrence, make, level, attention, question, clear, different*
10	*law, rights, human, convention, protection, treaty, criminal, space, weapons, against, sea, act, jurisdiction, outer, arms, armed, humanitarian, territorial, china, islands*
11	*electronic, north, south, korea, privacy, law, cyberspace, legal, korea's, regulations, health, surveillance, rules, concepts, including, traditional, present, ethics, difficult, companies*
12	*theory, cyberspace, decision, making, decisions, policy, process, been, using, was, explain, research, our, however, analysis, problem, concepts, key, model, issues*
13	*what, cyberterrorism, second, three, questions, first, two, analysis, traditional, between, threat, into, current, then, debate, concludes, strategies, literature, particularly, research*
14	*attacks, against, terrorism, critical, infrastructure, terrorist, groups, actions, different, various, networks, recent, other, attack, computers, components, also, part, infrastructures, conflicts*
15	*china, global, cooperation, united, power, economic, issues, policy, governance, china's, chinese, cyberspace, foreign, government, countries, among, order, context, russia, between*
16	*operations, capabilities, military, network, offensive, capability, support, approach, operational, use, defensive, terms, cyberspace, than, ability, concepts, defence, forces, attack, create*
17	*cybersecurity, development, strategic, also, city, countries, societies, smart, national, technology, need, through, resilience, yet, become, most, one, been, out, world*
18	*terrorist, use, world, legal, online, organizations, they, those, terrorists, who, activities, even, provides, through, violent, violence, all, governments, some, cyberspace*
19	*network, management, access, due, data, several, distributed, ddos, control, malicious, method, key, many, applications, users, based, system, services, number, service*
20	*data, systems, system, detection, intelligence, our, using, used, was, results, research, within, sources, intrusion, technologies, threat, knowledge, networked, process, time*
21	*public, media, social, between, democracy, governance, citizens, opinion, government, they, communication, study, how, show, was, content, were, regimes, about, authoritarian*
22	*politics, human, social, world, power, how, theory, virtual, environmental, what, movement, diplomacy, life, one, transnational, ways, real, about, relations, globalization*
23	*nuclear, most, many, two, important, but, between, weapons, when, what, become, state, into, yet, other, one, actors, against, iran, threat*
24	*attacks, attack, also, attacker, damage, traditional, future, power, particular, infrastructure, scenario, against, possible, cyberattacks, field, few, case, domain, platform, target*
25	*military, war, force, strategic, conflict, national, traditional, defense, e.g, than, cyberwar, cultural, concept, been, but, appropriate, resources, defend, kinetic, one*
26	*digital, crime, nato, science, prevention, forensic, region, including, strategic, legal, other, environment, through, part, national, challenges, area, energy, cybercrime, specific*
27	*malware, how, terrorism, used, organizations, analyze, use, software, problem, known, analysis, also, terrorist, set, better, files, about, there, structure, organizational*
28	*war, society, russian, world, what, 2007, russia, propaganda, conflict, use, western, ukraine, since, hybrid, against, military, order, conventional, itself, around*
29	*between, participation, study, public, activities, sphere, both, significant, online, culture, found, there, relationship, were, sample, procedural, findings, participants, effects, justice*
30	*technologies, technology, networks, military, over, but, may, computer, much, global, time, network, communications, than, less, only, term, control, loss, could*

Table 4. Results for 30-topic LDA model obtained from nearly 600 articles constituting the body of work in the cyber security and conflict field.

In order to assess the cyber conflict field and the place of terrorism and political violence language within it in a more granular fashion, Table 4 presents the results of a 30-topic LDA model obtained from nearly 600 articles constituting the body of work in the cyber security and conflict field. Again, though a large number of topic model sizes were explored, 12-topic and 30-topic models were chosen for presentation purposes as the correct standard for capturing different levels of semantic meaningfulness. Remarkably, the 30-topic model presents a relatively cohesive set of thematic research areas in the literature on cyber conflict. While there remain a broad number of topics constituted of relatively general terminology, there are distinguishing features between each. Several are clearly defined by empirical prompts in cyber conflict studies, such as the 2007 Russia-Estonia conflict,[67] the Olympic Games campaign against Iran's nuclear weapons development program,68 and cyber tensions on the Korean peninsula.[69] In essence, these areas of the field are common case studies that dominate in research. Some clearly suggest the shape of work on global Internet governance (and governance disagreements), while yet others outline important organizational (i.e., the development of NATO cyber forensics and defensive capacity) and technical topics.

Where the relatively granular breakdown of the terrorism studies field in the above section included a very limited interface with ICT-defined themes, the breakdown of the cyber field includes four topics that

---

[67] See P. Finn, "Cyber Assaults on Estonia Typify a New Battle Tactic," *Washington Post* May 19, 2007; R. Ottis, "Analysis of the 2007 Cyberattacks against Estonia from the Information Warfare Perspective," in *Proceedings of the 7th European Conference on Information Warfare*, 2008, 163; and A. L. Russell, *Cyber Blockades*, Washington, DC: Georgetown University Press, 2014.

[68] See A. Matrosov et al., "Stuxnet under the Microscope," *ESET LLC*, September 2010; J. R. Lindsay, "Stuxnet and the Limits of Cyber Warfare," *Security Studies* 22, no. 3 (2013): 365–404; D. Albright et al., *Did Stuxnet Take Out 1,000 Centrifuges at the Natanz Enrichment Plant?* Institute for Science and International Security, 2010; R. Langner, "To Kill a Centrifuge: A Technical Analysis of What Stuxnet's Creators Tried to Achieve," Hamburg: Langner Group, 2013; and K. Zetter, "How Digital Detectives Deciphered Stuxnet, the Most Menacing Malware in History," *Wired Threat Level Blog*, July 11, 2011, http://www.wired.com/ threatlevel/2011/07/how- digital- detectives-deciphered-stuxnet.

[69] See *inter alia* J. Jun et al., *North Korea's Cyber Operations: Strategy and Responses*, Rowman & Littlefield, 2016; and C. Whyte, "Ending Cyber Coercion: Computer Network Attack, Exploitation, and the Case of North Korea," *Comparative Strategy*, 2016.

are distinct from one another and focused on terrorism/political violence. Topics 13 and 14 suggest that shape of research engaging, respectively, with definitional questions on the nature of cyberterrorism and questions of the potential impact of cyberterrorist activities. Topic 18 appears to deal with terrorism from a law and governance perspective, with both the keyword list presented in Table 4 and the extended list (available in the appendix) relating the terrorist enterprise and terrorists' use of the web for organizational purposes to government oversight. Finally, Topic 27 appears to more directly deal with the integration of cyber tools and techniques into terrorist organizations.

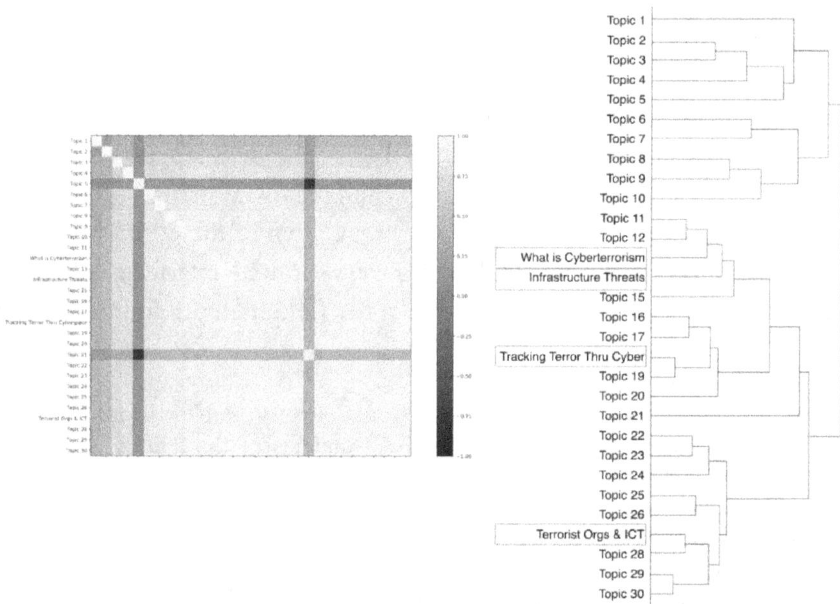

Figure 6. Correlation matrix and dendrogram visualization for Table 4's 30-topic LDA model.

Even without further visualization, the expanded model of cyber conflict research suggests that topics at the intersection of ICT and terrorism studies already constitute a distinct set of research agendas within the field. The question, however, remains as to the degree to which such themes are present in other parts of the cyber conflict field. Indeed, given that two of the terrorist-oriented topics appear to describe issues broadly relevant to other cyber research efforts in some granularity (i.e., the incorporation of new techniques into legacy organizational structures and the legal-governance basis of a form of cyber conflict), one might expect these topics to co-occur broadly across the field in thematically significant ways.

As Figure 6 demonstrates, however, there is mixed evidence to support such a hypothesis. It is certainly the case that each digital topic co-occurs broadly in much the way the lone "digital issues" topic did in the terrorism studies literature. Much as was the case in the section above, however, there is no evidence of significant major similarity with other themes in the correlation analysis. The story is somewhat different as told by the agglomerative hierarchical clustering visualization on the right in Figure 6. Here, it appears that the definitional topics (i.e., what is cyberterrorism and how impactful might it end up being?) are reasonably similar to a number of other major themes. While the topics appear in small proportions frequently across the entire corpus, it also seems to be the case that they share vocabulary with a number of other topics. This is perhaps unsurprising given that Topics 13 and 14 deal with definitional issues being grappled with across the entire cyber conflict field of study, from the prospective risk nature of different modes of cyber threat to the applicability of different conceptual frameworks. Topics 18 and 27, on the other hand, exhibit distinctly fewer topical similarities with other areas of the cyber conflict field. Much as was the case with the "digital issues" element of the terrorism studies field, these topics are empirically distinct. And, much as was the case with the terrorism studies literature, these empirically distinct topics present as a distinct sub-theme of the literature that are referenced in small part across a wide swathe of research works but that have few major thematic connections to other topics.

The Scope and Impact of Digital Research in the Field

Clearly, the terrorism studies and cyber conflict bodies of work assessed above constitute, when taken together, the broad horizon of research that might analytically link the information revolution with distinct topics in terrorism, political violence, and non-state conflict. What the analysis above does not speak to, however, is the temporal nature of thematic developments in the broader combined fields of cyber conflict and terrorism studies. Though modern information technologies have their roots in innovations made in the 1960s and cyber conflict has occurred for at least three decades, the global expansion and continued adoption of ICTs at the level of fundamental societal functions has naturally had a cumulative motivational effect on scholars of world affairs. In other words, many scholars have only recently turned to consider the effects of the information revolution on traditional security topics, on the one hand, and the development of new and distinct socio-technical phenomena, on the other. This section thus turns to an alternative form of topic modeling, dynamic topic modeling, in order to consider topical development in scholarship over time. In doing so, we are qualitatively presented with the opportunity to adjudicate on apparent propellants of new topic formation in scholarship.

Dynamic topic modeling alters a simple assumption made in the processing of obtaining traditional topic models, essentially that relationships between documents don't matter. This is a reasonable assumption from various points of view, as many inter-document relationships are relatively meaningless for the task of discovering latent thematic trends in a corpus. If a researcher wants to better understand topic formation around particular characteristics of individual documents, they may undertake semi-supervised topic modeling (as is done in the section below this) to note a particular unifying attribute of *some* documents over others. However, the order of documents still does not matter.

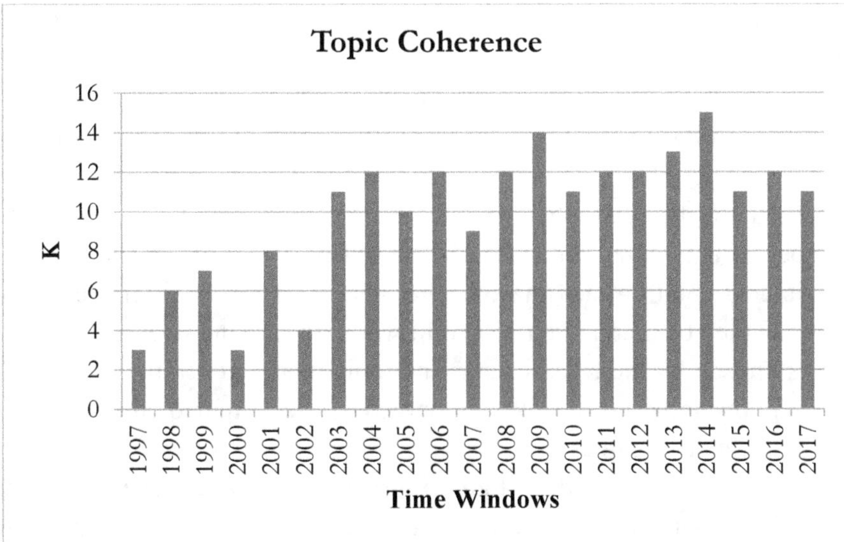

Figure 7. Topic coherence outputs of a Skipgram Word2Vec model obtained from a corpus of nearly 5,000 article abstracts drawn from Web of Science.

Naturally, any attempt to consider latent themes over time benefits from an ability to reference the order in which documents appear. Dynamic topic modeling allows researchers to do just that by obtaining topics in a two-step process. First, a series of topic models are produced based on the order of documents in a given corpus. These "window" topics, when the aim is to look at topics over time, reflect thematic trends within a given period of time, such as months, years, or decades. They also constitute the set vocabulary via which a broader set of dynamic topics can be generated that describes macro topics for the entire period of time under study. The result is an ability to study the formation of major themes in a corpus via dissection of contributing topics that present from time period to time period.

Figure 7 above presents the topic coherence outputs of a skipgram Word2Vec model[70] obtained from the combined corpus of nearly 5,000

---

[70] For description of the approach, see Winnie Cheng, Chris Greaves, and Martin Warren. "From n-gram to skipgram to concgram." *International journal of corpus linguistics* 11, no. 4 (2006): 411-433; and David Guthrie, Ben Allison, Wei Liu, Louise Guthrie, and Yorick Wilks. "A closer look at skip-gram modelling."

article abstracts drawn from Web of Science related to terrorism studies and cyber conflict. Though not necessary, it is possible to relinquish further control over the parameter specifications needed to obtain topic models with dynamic topic modeling. In producing a skipgram model, it is possible to obtain a measure of topic coherence for each window specified by the researcher. In essence, by setting broad limits on the number of topics that the researcher might be interested in finding, it is possible to obtain a measure of cohesiveness of different term clusters. This can then be used in lieu of an otherwise arbitrary number of requested topics chosen by the researcher in the topic model generation process at the window topic stage. For our purposes, the output of the skipgram word2vec model is interesting because it signals the diversification of the combined literature under study. In particular, topic coherence (i.e., identifiable, semantically meaningful topic areas in the field) appears to have boomed in the early 2000s.

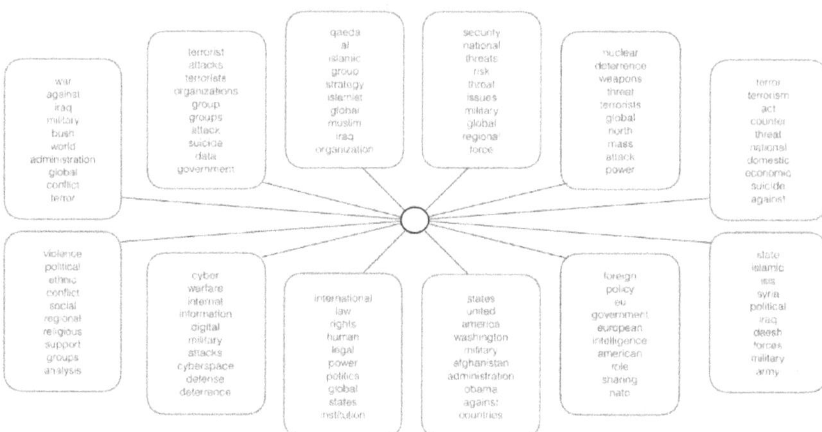

Figure 8. Results for a 12-topic non-negative matrix factorization dynamic topic model obtained from nearly 5,000 article abstracts drawn from Web of Science.

Figure 8 arrays the dynamic topics produced via initial assessment of window topics across a 21-year period from 1997 to 2017 (prior to

In *Proceedings of the 5th international Conference on Language Resources and Evaluation (LREC-2006)*, pp. 1-4, 2006.

1997, literature did not present as topically coherent enough to model with the dynamic approach). There exist a clear set of distinct issue areas. These include: (1) the War on Terror, (2) the shape of terrorism, (3) global jihadism and radicalization, (4) militaries and the use of force, (5) nuclear proliferation and terrorism, (6) counterterrorism, (7) ethnic violence and conflict, (8) cyber conflict and warfare, (9) international law and humanitarian issues, (10) American operations in Iraq and Afghanistan, (11) transatlantic security coordination, and (12) the Syrian/Islamic State conflicts. Naturally, for the purposes of this study, Topic 8, cyber conflict and warfare, is of prime interest.

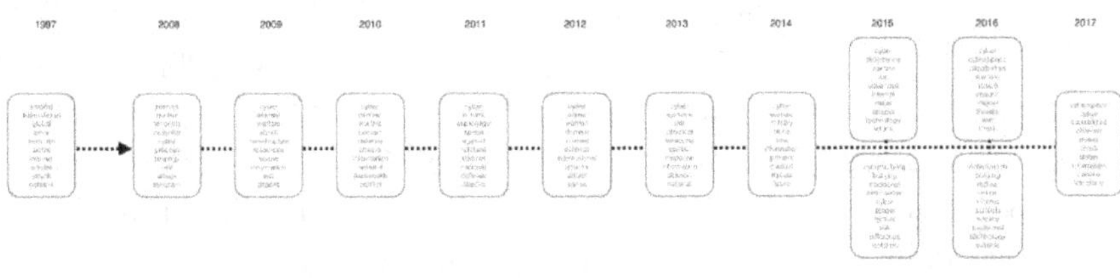

Figure 9. Window topic results for Topic 8 of Figure 8's 12-topic non-negative matrix factorization dynamic topic model.

Topic 8 quite naturally has its roots in the corpus in initial work on terrorism, broadly construed. This is unsurprising from two potential angles. First, the corpus of documents includes subject matter dominated by terrorism studies in the late 1990s and diversification of the field in the 2000s was largely driven by the events of the September 11th attacks on the United States. Second, early discussions of cyber conflict, or “netwar,” in the late 1990s were often linked with the possibility of terrorist attack.[71] This was the case for a few reasons, primarily that Clinton-era efforts to address cyber security had their roots in non-state and semi-state security incidents. These included the Oklahoma City bombings that defined Clinton's early presidency and the later events of Moonlight Maze and

[71] For narrative descriptions of the period, see *inter alia* R. A. Clarke and R. K. Knake. *Cyber War*, Tantor Media, Incorporated, 2014; and Jason Healey, ed. *A Fierce Domain: Conflict in Cyberspace, 1986 to 2012*, Cyber Conflict Studies Association, 2013.

SOLAR SUNRISE[72] (wherein the United States government was the target of espionage operations, respectively, by state-linked and independent hackers) that drove policymakers to act on cyber issues.

Interestingly, diversification of the cyber conflict topic does not become distinct until 2008 when the language of cyber warfare studies appears to diverge (given the systematic absence of previously prominent keywords) from that of terrorism studies. In the period between 2008 and 2014, window topics most closely linked with the overall dynamic topic illustrate more distinct thinking on the nature of cyberspace as a unique domain akin and related to space. In 2015, the dynamic topic diverges insofar as it is most closely linked with distinct topics focusing on cyber conflict and cyber security as a social issue. Though the appearance of keywords like deterrence do suggest a diversification of research focus within the cyber security field in the social sciences, the differentiation made between such thematic work and focus on cyberbullying and victimization illustrates a lack of diversity among cohesive topics in what IR scholars would traditionally consider digital security issues. More to the point here, the cyber conflict field itself does not present as interfacing with terrorism and political violence studies terms following early ruminations.

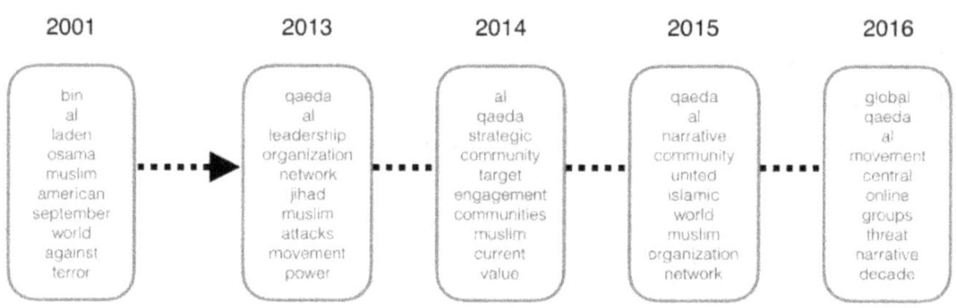

Figure 10. Window topic results for Topic 3 of Figure 8's 12-topic non-negative matrix factorization dynamic topic model.

Two other topics include keywords that smack of digital issues moving into the core of distinct thematic issue areas. Figure 10

[72] For perhaps the best in-depth descriptions of these events see F. Kaplan, *Dark Territory: The Secret History of Cyber War*. New York: Simon and Schuster, 2016.

demonstrates how window topics linked with the broader dynamic topic describing global jihad have shifted in focus over time in exhibiting community over organizational nomenclature. Where topics through 2013 rarely vary in key term usage beyond keywords like attacks, terrorize, organization, transnational, and network, core window topics from 2014 onwards reflect a shift in scholarly discourse focused on jihadi narrative and radicalization among global Muslim communities. In 2016, the word online was uniquely linked to the global jihad window topic, while the words digital, forum, and website appear in more expansive top twenty word lists in 2015 and 2016.

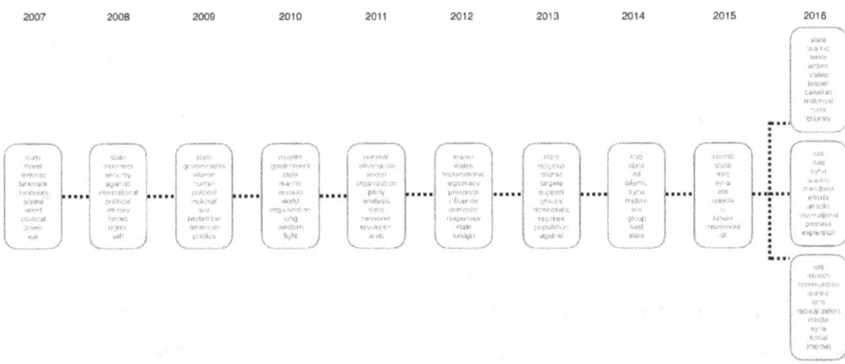

Figure 11. Window topic results for Topic 12 of Figure 8's 12-topic non-negative matrix factorization dynamic topic model.

Likewise, as Figure 11 shows, thematic rhetoric in the dynamic topic describing the Syrian and Islamic State conflicts has had a unique evolutionary trajectory from broader thematic topics that simply describe the shape of the terrorist enterprise. The dynamic topic appears to be defined by the intersection of discourse on specific conflicts, including those wars and the Arab Spring, and tensions between the Islamic and Western worlds more generally. In 2016, several window topics contributed to the construction of the dynamic topic, seemingly describing the threefold focus on the Islamic State as a territorial adversary, a terrorist organization, and a community element in the Muslim world. With this last topic, the interlinkages between the most recent Syria-Iraq set of conflicts and Muslim communities via digital media are quite apparent.

Discussion

Topic modeling presents a unique opportunity for assessing the latent thematic content of large bodies of scholarly work. In presenting a range of evidence derived from diverse application of topic models to scholarship on terrorism and cyber conflict studies, this article demonstrates that the cyber terrorism subfield (spread across the fields of cyber conflict and political violence studies) is *both* nuanced and limited. This section discusses a range of implications of the evidence outlined above.

A lack of conceptual precision

Broadly, this study supports and illustrates past arguments that the field, beyond one or two specific issue areas, remains topically imprecise in several ways. First, the use of the term "cyberterrorism" (and variants) is linked with remarkably little in the way of thematic distinction across different fields of study. Perhaps most interestingly, there is a clear lack of analytic and technical language in the topics resulting from analysis of the "cyberterror" (by keyword) corpus. Whereas we might expect to see distinction in content focusing on the tools of cyberterrorism as either being ancillary or linked with major infrastructural threats to national security, little beyond general terminology defines most topics produced. This suggests that reference to cyberterrorism is often a qualifying crutch in research more broadly on cyber or terrorism issues and that foundational definitional debate remains the bread and butter of much fuller scholarly engagement. Second, ICT wordage appears only sparingly in high-level topic model results obtained from both the cyber conflict and terrorism bodies of work. Where it does appear, such wordage is found alongside other non-traditional terminology, such as "biological." This strongly indicates that the state of the field is yet one of relative novelty.

Targeted impact, common reference

More specifically, distinct semantic themes that describe digital issues areas of study in the literature are referenced in small part across a wide swathe of research work but tend to have few major thematic connections to other topics. This supports the narrative that cyberterrorism, cyber-jihad and various other digital issue terminologies are referenced as a qualifier often in research across the broader field(s) without meaningful conceptual engagement. This is not meant as a critique, as a central criticism of scholars focused on cyber terrorism subject matter is that such a lack of conceptual precision and consensus exists. Nevertheless, it seems clear that such a shortcoming has manifested in limited engagement by scholars focused on other parts of the terrorist enterprise or other conflict issues. "Cyber security" and Internet-based terror are of unique interest, but they are analytically difficult to break apart and problematize.

Event driven

The field appears in analysis to be limited in another manner. Topic diversity often meets semantic meaningfulness in a topic's primary reference to specific events and conflicts. This is not surprising, to some degree, and even the broader terrorism studies takes reference at the highest level from defining events in recent history, such as the events of September 11th, 2001, and the resulting War on Terror. With cyber terrorism subject matter, however, granular assessment of the field reveals that disciplinarily unique themes (Islamic State military operations versus radicalization processes versus transnational terrorist network logistics, for instance) reference just a few conflicts. Specifically, much work in the field seems to focus on Israeli-Palestinian tensions, Islamic State and the Syrian conflict, and related terrorist activities around the world. While, again, it is not surprising that scholars might focus on such prominent subject matter, the focus on just a few case issues belies the potential scope of cyber terrorist activities and further reinforces the notion that conceptual divisions in the field over definitions present a barrier to empirical investigation of cyber antagonism by nonstate actors in general.

Missing techno-analytic engagement

In observing the results in sections above with an eye to what is missing, it seems strange that there is such a lack of analytic and technical language in topics. Specifically, it is unusual to see topic model results obtained from a corpus that represents a relatively cohesive subfield of study that contain no discrete analytic themes. Just as the topic model of the broader terrorism field above reflect several categories that describe the tools and terminology of scholars' approach to research, we might expect such terminology, whether technical or methodological, to appear among digital issues when bound as a subfield. This is not the case. Specifically, there appears to be almost no generalized inclusion of terminology linked with computer network attack or defense, such as "ddos," "malware," "rootkit," "virus," etc. While there is minimal evidence of such in the topic models produced from the narrow "cyberterrorism" keyword corpus, the same is not true for the broader field. Moreover, qualitative analysis of the "cyberterrorism" corpus suggests that such technical assessment occurs outside of the social sciences, where threat of digital terrorism is used to qualify the security implications thereof.

Missing topical focus

Beyond missing techno-analytic language, there are clear topical absences in the latent thematic areas defined in the figures above. In particular, there is surprising omission of content that might fall within the political violence field. Engagements on topics in digital activism and hacktivism, for instance, are curiously absent as themes in the broader corpus despite an arguable conceptual link with "cyberterrorism." After all, cyberattacks are not inherently violent, even if aggressive,[73] and a broad set of works in the cyber conflict field have noted the conceptual similarities between digital terrorism and other forms of coercion online, like hacktivism. It seems reasonable to assert that the continued expansion and deepening of work in the research program on nonstate cyber conflict would thus benefit from new focus on topics not seen in the results above, such as (among others):

- Cyber threats to civil society organizations (CSOs);
- Nonstate CSO cyber conflict, including:
  - Cyber-enabled subversion and propaganda;
  - Cyber coercion by non-terrorist actors;

[73] For those who argue that cyber weaponry is nonviolent in nature, see *inter alia* T. Rid, "Cyber War Will Not Take Place," *Journal of Strategic Studies* 35, no. 1 (2012): 5–32; and E. Gartzke, "The Myth of Cyberwar: Bringing War in Cyberspace Back Down to Earth," *International Security* 38, no. 2 (2013): 41–73.

  - Cyber-augmented criminal advocacy (e.g. "naming and shaming" operations undertaken by social movements).
- Foreign-focused cyber mercenary operations;
- The use of ICT in state sponsorship of terrorism;
- Augmentation effects of ICT usage alongside traditional terrorist acts.

Moreover, scholars would do well to employ theoretical frameworks common to the political violence field in engaging on such topics. In many cases, it is likely that the above topics are not understudied or completely absent. Rather, they are found as objects of study in other academic fields where there is simply no thematic engagement with work in the terrorism studies field.

Conclusion

This article has provided empirical support for recent arguments that the focus on the intersection of digital and terrorism studies lacks analytic precision and diversity on several fronts. Given the evidence above, it seems reasonable to suggest that continued definitional squabbles and focus hamper the development of research on the Internet, political violence, radicalization, and cyber conflict. As such, I echo the call made by other scholars in this domain over the past two years to redouble efforts to add conceptual nuance to the field enabled by smarter development of empirical data collection and testing approaches.

Specifically, I argue that the best approach for achieving such conceptual and empirical diversity would be the adoption of a program of micro-foundational compartmentalization at the project level. By adopting less restrictive conceptual assumptions about the traditional bounds of extremism, political violence, and the terrorist enterprise, scholars can work to develop new understanding of how ICTs have impacted threat actors and processes. In particular, scholars should expand the range of activities that might be considered to constitute political violence or radicalization efforts. In doing so, they will inevitably expand focus on actors clearly not often considered in contemporary efforts to problematize conflict in the digital age, including subversive organizations, hacktivist outfits, organized criminal entities, and non-jihadi extremists. Likewise, researchers would do well to augment existing methodological strengths with new data production schemes. The information revolution has not only been immensely impactful on the subjects of study for the terrorism study field; it has also provided a host of new tools for assessing and operationalizing antagonistic, conflictual behavior. Finally, scholars in this domain should actively look to cross interdisciplinary boundaries in considering the socio-technical design side of issues (from the design of malware to the shape of relevant parts of the Darkweb ecosystem) and collaborating with those in the computational sciences. In doing so, future assessments of the shape of this subfield will hopefully find a more diverse, expansive, and thematically precise set of research programs.

www.ingramcontent.com/pod-product-compliance
Lightning Source LLC
LaVergne TN
LVHW010533300926
849162LV00016B/1870/J
* 9 7 8 1 6 8 0 5 3 9 9 8 1 *